I0198071

HISTORICAL BIBLIOGRAPHY OF
SEA MINE WARFARE

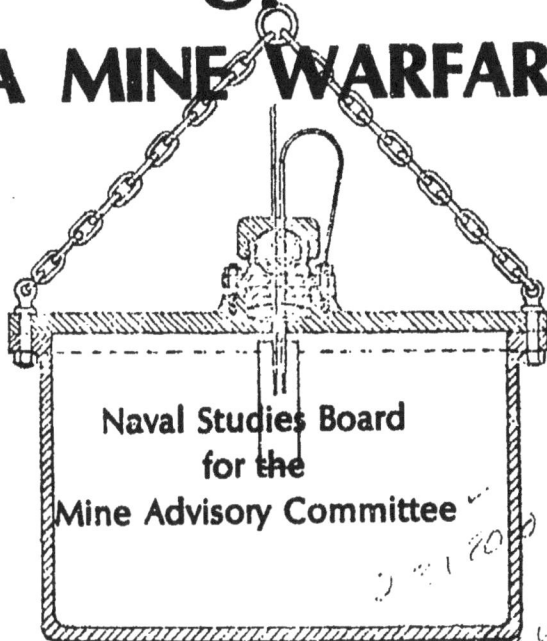

Naval Studies Board
for the
Mine Advisory Committee

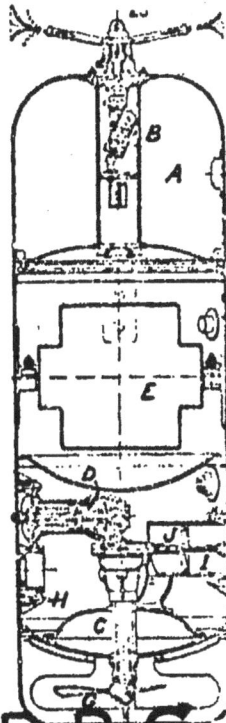

D D C

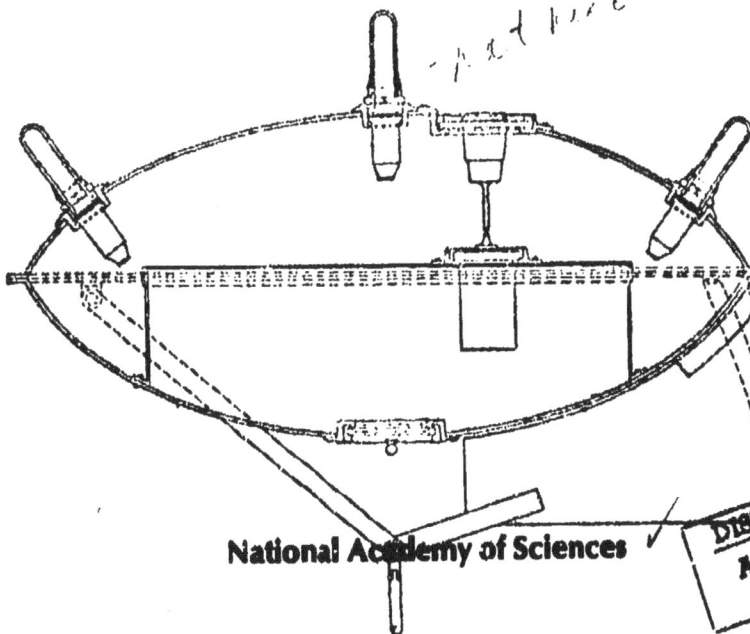

National Academy of Sciences

78 09 11 001

(14)

NRG-MAC-2033

(6)

HISTORICAL BIBLIOGRAPHY

OF

SEA MINE WARFARE

(10) Edited
by
Andrew Patterson, Jr
and
Robert A. Winters

(11) Jan 76

(12) 146p.

(15) N00014-75-C-0258,
N00014-67-A-0244-0011

Naval Studies Board
for the
Mine Advisory Committee
National Research Council

£473

National Academy of Sciences
1977

231800

78 09 11 001

AB

FOREWORD

On January 1, 1974, after 23 years of continuous service to the Navy and the Mine Warfare Community, the Mine Advisory Committee was replaced by the Naval Studies Board. At the termination of the Committee there remained a single piece of unfinished business--the compilation of a "Historical Bibliography of Sea Mine Warfare." The Board believes that such a collection of historical references will be of value to both naval historians and those scientists, engineers and naval officers who believe the past holds the key to the future. Accordingly, the editors--Dr. Andrew Patterson, Jr., and Mr. Robert A. Winters--were asked to complete their task. This volume is the result of their efforts.

The work on this volume was begun under Contract N00014-67-A-0244-0011 and completed under Contract N00014-75-C-0258 with the Office of Naval Research.

PREFACE

Early in 1968 the Mine Advisory Committee of the National Academy of Sciences-National Research Council undertook, at the request of the Navy, one of the more exhaustive of its many studies of sea mine and sea mine countermeasures technology. Entitled "The Present and Future Role of the Mine in Naval Warfare" (short title: PROJECT NIMROD), the study covered a period of two years, and resulted in a two-volume report dated January 15, 1970.

One of the several objectives of the PROJECT NIMROD effort was to correct a number of misconceptions regarding the effectiveness of the mine as a modern naval weapon: misconceptions which consistently rendered more difficult the support of an aggressive research and development program. Since these misconceptions had evolved over many years and were based largely on second and third hand impressions of past naval engagements, it was considered essential that the record be set straight. To accomplish this the Committee assigned the authors the task of compiling a brief but carefully documented history of the sea mine since its invention by David Bushnell, in 1775. The results of this effort were included in the final report of PROJECT NIMROD, and subsequently issued as a separate report entitled "A Brief History of Mine Warfare."

Research on the "History" provided an unique opportunity to assemble most of the historical literature on mine warfare. Having gone to considerable effort to identify and gain access to well over two thousand documents, continuing to the present time, we and the Committee believe the effort should be preserved in the form of a bibliography for use by future researchers. This document is offered in that spirit, and for that purpose. The authors will be pleased to have pointed out to them additional references to historical documents, and any errors (of which inevitably there will be some) which are discovered in this bibliography.

The authors have been persuaded it is more desirable to put this bibliography before the public now than to delay publication while providing it with various scholarly aids. This decision has been imposed, in part, by the necessity for all those involved in preparing it to do so as a part-time activity low on a list of priorities. To accomodate the restrictions under which we have worked, while yet making it possible to prepare later appendices which might include cross-references and subject, category, or temporal indices, we have numbered the entries on each page in order. Thus 2-11 will identify the eleventh entry on page two. The entries in the present version are alphabetical either by author (e.g., 2-11), by subject (2-5), by title (2-8), by pseudonym (2-1), etc.

Although in its present form the entries in the bibliography are not accompanied by interpretive comments, the authors have personally perused many of the works listed and have such comments in their files. We welcome the opportunity to exchange information about the listings with users of this document. Those interested can address their comments to us using the reference numbers described above. We hope feedback from users will assist us in preparing a second edition. Since the authors cannot have examined every entry, such user comments will be especially valuable.

The selected documents listed on the pages to follow were drawn from many sources, and with the able assistance of many people. Prominent among these sources were: The Library of Congress, Navy Historical Section; Department of the Navy, Naval History Division; Naval War College Library; Naval Ordnance Laboratory Library; Archives of the Torpedo Station, Newport, Rhode Island; Yale Sterling Memorial and Beinecke Libraries; Weidner Library at Harvard; Reading Room and Newspaper Room of the British Museum; Public Record Office (London); the Science Museum (London); Admiralty Library (London); the Library of the United Services Institution (London); Library of the Ministry of Defense (London); Library of the Royal Engineers (Chatham); the Bodleian Library and Oxford college libraries; the Bibliotheque de la Marine in Paris; Library of the Royal Danish Navy (Copenhagen); the Bibliotek für Zeitsgeschichte in Stuttgart, and the Bundesarchiv-Militärarchiv in Freiburg.

Mention should also be made of very productive visits with Vice Admiral (Ret) Kyuzo Tamura, who was in charge of the clearing of mines from Japanese waters after World War II, and for mine activities in various posts during the war, and with Rear Admiral (Ret) Friedrich Ruge, who was Rommel's Chief of Staff before D-Day, and who held various important mine warfare posts in the German Navy from World War I through World War II. Admiral Tamura and his staff produced documentation and information available from no other source, since all official documents were burned before the end of World War II. Admiral Ruge made available items from his personal files and library which are of the utmost value.

Finally, we should like to recognize, with deep appreciation, the secretarial staff of the Mine Advisory Committee and of the Naval Studies Board for their assistance in locating documents, for processing reams of document request forms, and for the preparation of the manuscript.

Andrew Patterson, Jr.

Robert A. Winters

Washington, D.C.

Mine Advisory Committee*
Membership

Chairman
Dr. James S. Coles
President, Research Corporation

Members
Dr. Arnold B. Arons
Department of Physics
University of Washington, Seattle

Dr. Hugh Bradner
Professor of Applied Physics and
  Geophysics
Institute of Geophysics and
  Planetary Physics
University of California, SD

Dr. Daniel B. DeBra
Professor of Aeronautics and
  Astronautics
Stanford University

Dr. R. Stuart Mackay
Department of Biology
Boston University

Dr. John J. McKetta
Graduate Professor of Chemical
  Engineering
University of Texas at Austin

Dr. Chester M. McKinney
Director, Applied Research
  Laboratories
University of Texas at Austin

Dean David Mintzer
Associate Dean
Technological Institute
Northwestern Institute

Dr. Arthur O. Williams, Jr.
Department of Physics
Brown University

Associate Members
Dr. Arthur L. Bennett
Research Professor of Physics
Engineering Experiment Station
Georgia Institute of Technology

Dr. Alfred B. Focke
Physics Department
Harvey Mudd College

Dr. Karl F. Herzfeld
Head, Department of Physics
Catholic University of America

Professor John D. Isaacs
Scripps Institution of Oceanography
University of California, SD

Dr. John A. Knauss
Graduate School of Oceanography
University of Rhode Island

Dr. Andrew Patterson, Jr.
Sterling Chemistry Laboratory
Yale University

Dr. Joseph B. Platt
President, Harvey Mudd College

Staff
Mr. Lee M. Hunt
Executive Secretary

Mr. Lester W. Barber
Professional Associate

Mr. Robert A. Winters
Professional Associate

---

*Terminated January 1974.

Naval Studies Board

Membership

Chairman
Dr. William H. Pickering
Professor of Electrical Engineering
California Institute of Technology

Mr. Benjamin B. Bauer
Vice President and General Manager
CBS Technology Center

Dr. D. Allan Bromley
Chairman, Department of Physics
Director, Wright Nuclear Structure
  Laboratory
Yale University

Dr. John C. Calhoun, Jr.
Vice President for Academic Affairs
Texas A & M University

Professor George F. Carrier
Division of Engineering and
  Applied Physics
Harvard University

Dr. Francis H. Clauser
Clark B. Millikan Professor of
  Engineering
California Institute of Technology

Engene P. Cronkite, M.D.
Chairman, Medical Department
Medical Research Center
Brookhaven National Laboratory

Dr. Willis H. Flygare
School of Chemical Sciences
University of Illinois

Dr. Ivan A. Getting
President, The Aerospace Corporation

Dr. John A. Hornbeck
Vice President
Bell Laboratories

Dr. Klaus E. Knorr
Professor of Politics and
  International Affairs
Woodrow Wilson School
Princeton University

Dr. A. Robert Kuhlthau
Department of Engineering Science
  and Systems
School of Engineering and Applied
  Sciences
University of Virginia

Dr. Vincent V. McRae
Director, Strategic Planning
IBM Corporation

Dr. Joseph M. Reynolds
Vice President, Instruction and Research
Louisiana State University

Dr. Harrison Shull
Vice Chancellor for Research and
  Development
Indiana University

Staff Officer
Lee M. Hunt

Abbot, Henry Larcom; *Army and Navy Journal*; Vol. XLI, No. 52, Aug. 27, 1904;   1.
New York. pg. 1344.

Abbot, Henry Larcom; *The Beginning of Modern Submarine Warfare Under*   2.
Captain-Lieutenant David Bushnell; Professional Paper No. 3, Engineer
School; Battalion Press, 1881; Willets Point, New York.

Abbot, Henry Larcom; *The Beginning of Modern Submarine Warfare: Under*   3.
Captain and Lieutenant David Bushnell, Sappers and Miners, Army of
the Revolution; Battalion Press, Corps of Engineers, 1881; Willets
Point, New York.

Abbot, Henry Larcom; *Early Days of the Engineer School of Application*;   4.
Occasional Papers, No. 14, Engineer School of Application; U.S.
Army, Washington Barracks, 1904; Washington, D.C. [Bound with
Papers No. 8-14.]

Abbot, Henry Larcom; *National Cyclopedia of American Biography*; Vol. XI,   5.
James T. White and Co., 1909; New York. pp. 194-195.

Abbot, Henry Larcom; *Our Torpedo System*; Army and Navy Journal, Vol. XVII,   6.
No. 24, Jan. 17, 1880; New York. pp. 465-466.

Abbot, Henry Larcom; *Professional Papers Corps of Engineers, U.S.A., No.*   7.
23 (Submarine Mines); Battalion Press, Corps of Engineers, 1881;
Willets Point, New York.

Abbot, Henry Larcom; *Report Upon Experiments and Investigations to Develop*   8.
A System of Submarine Mines for Defending the Harbors of the U.S.;
Engineer Department, War Department; U.S. Government Printing Office,
1881; Washington, D.C.

Abbot, Henry Larcom; *Torpedoes for Harbor Defense*; Army and Navy Journal,   9.
Vol. XI, No. 25, Jan. 31, 1874; New York. pg. 392.

Abbot, Henry Larcom; *Torpedoes for Harbor Defense*; Army and Navy Journal,   10.
Vol. XI, No. 26, Feb. 7, 1874; New York. pg. 410.

Abel, F. A.; *Electricity Applied to Explosive Purposes*; The Institution   11.
of Civil Engineers, 1883. [A separately bound pamphlet.]

Adams, William T.; *The Underwater War*; Ordnance, Vol. XLVII, No. 255,   12.
Nov-Dec. 1962; Washington, D.C. pp. 317-319.

"Admiral" (column Writer's By-Line); United Service Magazine, Vol. L, No.   13.
1036, Mar. 1915; London. pg. 571.

1.   "Admiral"; *The Navy and the War*; United Service Magazine, Vol. L,
       No. 1032, Nov. 1914; London. pp. 117-124.

2.   *The Admiralty Torpedo School*; Engineer, Vol. XLII, Dec. 29, 1876;
       London. pg. 446.

3.   *H.M.S. Adventure*; Professional Notes, U.S. Naval Institute Proceedings,
       Vol. LII, No. 8, Aug. 1926; Annapolis. pg. 1585.

4.   *H.M.S. Adventure*; Professional Notes, U.S. Naval Institute Proceedings,
       Vol. LIII, No. 2, Feb. 1927; Annapolis. pg. 349.

5.   (Aerial Mines); L'Europe Nouvelle, Vol. XXII, Apr. 29, 1939; Paris.
       pg. 460.

6.   *Aerial Mine Sweeper*; Newsweek, Vol. XXI, No. 4, Jan. 25, 1943; Washington,
       D.C. pg. 19.

7.   *Aerial Tugs*; Military Review, Vol. XXXV, Jun. 1955; Ft. Leavenworth.
       pg. 64.

8.   *Air Ministry Account of the Part Played by the Coastal Command in the
       Battle of the Seas, 1939-1942*; MacMillan, 1943; New York.

9.   Albert, Prince de Monaco; *Les Mines Errantes sur l'Atlantic Nord*;
       Comptes rendus des séances de l'Academie des Sciences, Vol. CLXX,
       No. 13, Mar. 29, 1920; Paris. pp. 778-782.

10. Albion, Robert Greenhalgh and Jennie Barnes Pope; *Sea Lanes in
       Wartime - The American Experience 1775-1945*; 2nd Ed., Anchor
       Books, 1968.

11. Alden, Carroll S. and Allan Westcott; *The United States Navy: A
       History*; Lippincott Co., 1943; New York.

12. Alden, John D.; *Flush Decks and Four Pipes*; U.S. Naval Institute, 1965;
       Annapolis.

13. Alexander, Ray; *The Cruise of the Raider, Wolf*; Yale University
       Press; 1939.

14. Alexinsky, Gregor; *Russia and the Great War*; T. Fisher Unwin, 1915;
       London.

15. Alger, Philip R., Jr.; *The Employment of Submarine Mines in Future Naval
       Wars*; U.S. Naval Institute Proceedings, Vol. XXXIV, No. 3, Sept.
       1908; Annapolis. pp. 1039-1042.

L'Allemagne Fait la Guerre au Droit des Gens; L'Europe Nouvelle, Vol. XXII, 1.
    Dec. 2, 1939; Paris. pp. 1338-1340.

Alliman, A. L.; *Torpedoes;* Engineer, Vol. XLIV, 1877; London. pg. 367. 2.

Allison, R.; *Torpedoes;* Engineer, Vol. XLII, 1877; London. pg. 446. 3.

H.M.S. Althom in Thames; British Motor Ship, Vol. XXXVI, No. 426, 4.
    Sept. 1955; London. pp. 254-256. [Wood-Aluminum Minesweeper.]

Altogether; Army and Navy Journal, Vol. XX, No. 13, Oct. 28, 1882; 5.
    New York. pg. 293.

The American Naval Planning Section London; U.S. Government Printing Office, 6.
    1923; Washington, D.C.

American Torpedoes in England; Scientific American, Vol. XIII, No. 22, 7.
    Nov. 25, 1865; New York. pg. 350.

Ammen, Daniel; *The Navy in the Civil War: The Atlantic Coast;* Sampson 8.
    Low, Marston, 1898; London.

Amphibious Warfare: Navy Air Force Commandoes; The Times Literary Supplement, 9.
    No. 2155, Saturday, May 22, 1943; London. pg. 242.

Anderson, Bern; *By Sea and By River, The Naval History of the Civil War;* 10.
    Knopf, 1962; New York.

Anderson, Frank; *Beginning of Modern Submarine Warfare Under Captain* 11.
    Lieutenant David Bushnell; by Henry L. Abbot, Anchor Books,
    1966; Hamden, Connecticut.

Anderson, Jack; *McCain's Swan Song;* Washington Post, Vol. XCV, No. 309, 12.
    Monday, Oct. 9, 1972; Washington, D.C. pg. D-15.

Anderson, Jane and Gordon Bruce; *Flying, Submarining and Mine Sweeping;* 13.
    Sir Joseph Causton & Sons, Ltd., 1916.

Andrews, A. F. and P. R. Hoang; *Mine Sweeper Computer Simulation;* U.S. 14.
    Naval Postgraduate School - Research Paper No. 64, Feb. 1966. 55 p.

Annual Reports H.M.S. Vernon. [Indices list many mine papers.] 15.

Annual Reports of the War Department for the Fiscal Year Ended June 30, 16.
    1903, Vol. VI, Military Schools and Colleges, etc; Submarine Defense
    School, Ft. Totten. U.S. Government Printing Office, 1903; Washington,
    D.C.

1. Ansel, Walter; *Hitler Confronts England*; Duke University Press, 1960;
    Durham, North Carolina.

2. Anthony, Irvin; *Decatur*; Charles Scribners Sons, 1931; New York.

3. Appleton, C.; *Appleton's Cyclopedia of American Biography*; Vol. IV,
    LOD to PICK, D. Appleton Co., 1894; New York. pp. 264-266.
    [M. F. Maury]

4. Armagnoe, A. P.; *Can Mines Conquer Sea Power?*; Popular Science,
    Vol. CXXXVI, No. 3, Mar. 1940; New York. pp. 78-83.

5. *Armament of the Minelayers*; Professional Notes, Vol. XXV, U.S. Naval
    Institute Proceedings, No. 3, Sept. 1909; Annapolis. pg. 965.

6. *L'Armée et la Flotte du Japon. Composition et Repartition en 1904*;
    Berger-Levrault & Cie., 1905; Paris.

7. *L'Armée et la Flotte de la Russie. Composition et Repartition en 1904;
    L'Armée de Mandchourie*; Berger-Levrault & Cie., 1905; Paris.

8. *Les Armées et les Flottes Militaires de tous les États de Monde. Composition
    et Repartition en 1904*; Berger-Levrault & Cie., 1905; Paris.

9. Armstrong, James; *The Sweeper of the Seas*; Navy and Army Illustrated,
    Vol. IV, New Series, 1915; London. pp. 24-25.

10. *Army Mine Planter Service*; Coast Artillery Journal, Vol. LXIV, No. 5,
    May 1926; Ft. Monroe, Virginia. pp. 526-530.

11. *Army Torpedo School*; Army and Navy Journal, Vol. XXXIX, No. 7, Oct. 19,
    1901; New York. pg. 151.

12. Artillerie navale; *Renseignements Divers à L'Usage des Officiers*;
    Imprimerie Nationale, 3e serie, Tome I, 1er fascicule, 1908; Paris.
    111 p.

13. *The Artillery and Other War Material Produced at the International Exhi-
    bition, Paris*; The Engineer, Vol. XXV, Jan. 3, 1868; London. pp. 1-2.

14. *The Artillery and Other War Material Produced at the International Exhi-
    bition, Paris*; The Engineer, Vol. XXV, Jan. 10, 1868; London. pp. 19-20.

15. *The Artillery and Other War Material Produced at the International Exhi-
    bition, Paris*; The Engineer, Vol. XXV, Jan. 17, 1868; London. pg. 35.

16. Ashbrook, Allan Withes; *Naval Mines*; U.S. Naval Institute Proceedings,
    Vol. XLIX, No. 2, Feb. 1923; Annapolis. pp. 303-312.

Ashley, L. R. N.; *The Royal Air Force and Sea Mining in World War II*; Air University Quarterly Review, Vol. XIV, No. 3, Summer 1963; Maxwell Field, Alabama. pp. 38-48.   1.

Ashton, George; *Mine Sweeping Made Easy*; U.S. Naval Institute Proceedings, Vol. LXXXVII, No. 7, Jul. 1961; Annapolis. pp. 66-71.   2.

Aspinall-Oglander, Cecil F.; *History of the Great War, Military Operations Gallipoli*; William Heinemann Ltd., Vol. I 1929; Vol. II 1932; London.   3.

Aspinall-Oglander, Cecil F.; *Roger Keyes*; (Being the biography of Admiral of the Fleet Lord Keyes of Zeebrugge and Dover), Hogarth Press, 1951; London.   4.

Assmann, Kurt; *German Preparations for Attacking the U.S.S.R.*; 1947. Pamphlet.   5.

Aston, George G.; *Letters on Amphibious Wars*; John Murry, 1911; London. pp. 283-284.   6.

ASW Search and Surveillance Program Advisory Board Summary Report (U); Naval Material Command, Nov. 10, 1972. SECRET-NOFORN.   7.

Atlantic Mine Force; All Hands, No. 584, Sept. 1965; Washington, D.C. pp. 6-7.   8.

At Sea: Ambitious Answer; Time, Vol. XXXV, No. 2, Jan. 8, 1940; New York. pg. 28.   9.

At Sea: Royal Navy's Test; Time, Vol. XXXV, No. 17, Apr. 22, 1940; New York. pp. 19, 20, 22.   10.

Attack and Defense by Submarine Mines; (The Most Easily Prepared and Most Dreaded Form of Naval Warfare), Scientific American, Vol. CXI, No. 14, Oct. 3, 1914; New York. pp. 270-271 and 286.   11.

Audic, M.; *Cours de Torpilles ...*; Imprimerie Nationale, 1880; Paris. 146 p.   12.

Audic, M.; *Étude sur les Effets des Explosions Sous-Marines*; Revue Maritime et Coloniale, Vol. LIV, Sept. 1877. pp. 531-601.   13.

Auphan, Paul and Jacques Mordal; *The French Navy in World War II*; U.S. Naval Institute, 1959; Annapolis, Maryland.   14.

The Austrian Torpedo; Van Nostrand's Engineering Magazine, Vol. II, No. 15, Mar. 1870; New York. pp. 325-326.   15.

An Automatic Mine; Professional Notes, Journal of the U.S. Artillery, Vol. XLIV, Jul-Dec. 1915; Ft. Monroe, Virginia. pp. 230-231.   16.

1. *Automatic Submarine Mines;* Engineering, Vol. XCIII, Apr. 19, 1912;
        London. pp. 520-524.

2. *Automatic Submarine Mines;* Engineering, Vol. XCIII, May 3, 1912;
        London. pp. 584-587.

3. *Automatic Submarine Mines;* Journal of U.S. Artillery, Vol. XXXVIII,
        No. 1, Jul-Aug. 1912; Ft. Monroe, Virginia. pp. 70-85.

4. *Automatic Submarine Mines;* Journal of U.S. Artillery, Vol. XXXVIII,
        No. 2, Sept-Oct. 1912; Ft. Monroe, Virginia. pp. 232-240.

5. *Auxiliary Vessels;* Professional Notes, U.S. Naval Institute Proceedings,
        Vol. XXVIII, No. 2, Jun. 1912; Annapolis, Maryland. pg. 775.

6. Avery, Ray L.; *The Mine Defense of Harbors: Its History, Principles,
        Relation to the Other Elements of Defense, and Tactical Employment;*
        Journal of the U.S. Artillery, Vol. XLII, Jul-Dec. 1914; Ft. Monroe,
        Virginia. pp. 1-17.

7. Avezathe, B.; *Mines in the Fairway! Gallant Fishermen of England;*
        Christian Science Monitor Magazine, Jan. 24, 1942; Boston.
        pp. 8-9.

8. *Baby Minesweepers - MSB's - Soon to Join Fleet;* All Hands, No. 433,
        Mar. 1953; Washington, D.C. pg. 5.

9. Bagby, Oliver W.; *Naval Mining and Naval Mines;* U.S. Naval Institute Pro-
        ceedings, Vol. LI, No. 12, Dec. 1925; Annapolis, Maryland. pp. 2244-
        2257.

10. Baggott, A. J. and C. H. Fawcett; *Development on Magnetic and Acoustic
        Mines at Admiralty Mining Establishment;* Journal of the Institution
        of Electrical Engineers, Vol. XCIV, Part I, No. 83, Nov. 1947;
        London. pp. 509-526.

11. Baggott, A. J. and C. H. Fawcett; *Discussion on "Development on Magnetic
        and Acoustic Mines at Admiralty Mining Establishment";* Journal of
        the Institution of Electrical Engineers, Vol. XCV, Part I, No. 96,
        Dec. 1948; London. pp. 550-551.

12. Baird, C. W.; *Employment of Mines in Coast Defense;* Coast Artillery
        Journal, Vol. LXXV, No. 4, Jul-Aug. 1932; Ft. Monroe, Virginia.
        pp. 260-261.

Baird, George Washington; *Additional Notes on Submarines*; Journal of 1.
American Society of Naval Engineers, Vol. XXVII, 1915; Washington,
D.C. pp. 186-191.

Baker, H. G.; *The Mine Force Wooden Ships and Iron Men*; All Hands, 2.
No. 481, Feb. 1957; Washington, D.C. pp. 6-9.

Barber, Francis Morgan; *Lecture on Movable Torpedoes*; U.S. Torpedo 3.
Station, Dec. 1874; Newport, Rhode Island. pp. 5-15.

Barber, Francis Morgan; *Professional Notes, The Destruction of the United* 4.
States Battleship - Maine; U.S. Naval Institute Proceedings, Vol.
XXV, No. 3, Mar. 1899; Annapolis, Maryland. pp. 515-516.

Barber, Francis Morgan; *The Progress of Torpedo Warfare*; The United 5.
Service Magazine, Vol. III, No. 3, Sept. 1880; Philadelphia,
Pennsylvania. pp. 278-293.

Barclay, Thomas; *Floating Mines Curse*; Living Age, Vol. CCLXXXIII, 6.
Dec. 19, 1914; Boston. pp. 715-720.

Barclay, Thomas; *The Floating Mines Curse, an Unsentimental Study*; 7.
The Nineteenth Century and After, Vol. LXXVI, No. 452,
Oct. 1914; London. pp. 745-752.

Barnard, Henry; *Armsmear, The Home, The Arm and the Armory of Samuel* 8.
Colt, A Memorial; Publisher unknown, 1866; New York.

Barnes, John Sanford; *Submarine Warfare*; Van Nostrand, 1869; New York. 9.

Barnes, Lt. Commander; *Les Torpilles. La Guerre Sous-Marine*; Imprimerie, 10.
P. Dupont; Paris. 42 p. [A partial translation of the above.]

Barnes, Lt. Commander; *Submarine Warfare Offensive and Defensive ...*; 11.
Van Nostrand, 1869; New York. 233 p. and figures.

Barrow, John; *An Autobiographical Memoir of Sir John Barrow, Bart., Late* 12.
of the Admiralty; Murray, 1847; London. pp. 274-275.

"Bartimeus"; *Sweeping Death's Doorstep*; Atlantic, Vol. CLXVII, 13.
Mar. 1941; Boston. pp. 288-294.

"Bartimeus"; *Mines-Hidden Peril of the Sea*; Sea Power, Vol. II, 14.
No. 4, Apr. 1942; Washington, D.C. pp. 10-11.

1. Bartlett, C. J.; *Great Britain and Sea Power;* 1963; Oxford.

2. Bartlett, C. J.; *Great Britain and Sea Power 1815-1853;* Clarendon Press, 1963; Oxford.

3. Bates, P. L.; *Naval Mines;* Military Review, Vol. XXXIII, No. 1, Apr. 1953; Ft. Leavenworth. pp. 48-56.

4. *Battle of the Atlantic - New Hazard;* Time, Vol. XXXIX, No. 26, Jun. 29, 1942; New York. pg. 26.

5. Baugh, Barney; *History of Naval Mine Warfare - Kegs, Cabbages and Acoustics;* All Hands, No. 481, 1957; Washington, D.C.

6. Beals, Victor; *New England's Role in the Yankee Mining Squadron;* The New England Galaxy, Vol. I, No. 1, Summer 1968; Boston. pp. 3-11.

7. (Beardslee's experiments); Nautical Magazine, Nov. 1865; pp. 601-603.

8. *Beaufighting;* Newsweek, Vol. XIX, No. 26, Jun. 1942; Washington, D.C. pp. 20-21.

9. Beehler, W. H. (Translator); *Scandinavian Experiments with Submarine Mines;* U.S. Naval Institute Proceedings, Vol. VII, No. 16, 1881; Annapolis, Maryland. pp. 121-154.

10. Belknap, Reginald Rowan; *The Barrage that Stopped the U-Boat I;* Scientific American, Vol. CXX, No. 11, Mar. 15, 1919; New York. pp. 250-251.

11. Belknap, Reginald Rowan; *The Barrage that Stopped the U-Boat II;* Scientific American, Vol. CXX, No. 12, Mar. 22, 1919; New York. pp. 288-289.

12. Belknap, Reginald Rowan; *North Sea Mine Barrage;* National Geographic, Vol. XXXV, Feb. 1919; Washington, D.C. pp. 85-110.

13. Belknap, Reginald Rowan; *Submarine Mines in War;* Journal of the U.S. Artillery, Vol. LI, No. 5, Nov. 1919; Ft. Monroe, Virginia. pp. 455-471.

14. Belknap, Reginald Rowan; *The Yankee Mining Squadron;* U.S. Naval Institute, 1920; Annapolis.

15. Belknap, Reginald Rowan; *The Yankee Mining Squadron;* U.S. Naval Institute Proceedings, Vol. XLV, No. 12, Dec. 1919; Annapolis, Maryland. pp. 1972-2012.

Bell, Archibald Colquhoun; *The Blockade of the Central Empires 1914-1918*; H.M.S.O., 1961; London.   1.

Bell, Archibald Colquhoun; *The Economic Blockade*; Her Majesty's Stationery Office, 1927, 1960; London.   2.

Bell, Archibald Colquhoun; *A History of the Blockade of Germany and of the Countries Associated with Her in the Great War*; Her Majesty's Stationery Office, 1937; London.   3.

Bell, Archibald Colquhoun; *Sea Power and the Next War*; Longmans, Green, 1938; London.   4.

Bell, L.; *Submarine Mines in Modern Warfare*; National Magazine, Vol. VIII, 1898; Boston. pg. 458.   5.

Bell, L.; *Submarine Mines in Modern Warfare*; Revue of Revues, Vol. XVIII, No. 2, Aug. 1898; New York. pp. 213-214.   6.

Bellet, Daniel; *Torpilles Mobiles, Torpilles Fixes Mines Sous-Marines*; Revue du mois, Vol. XIX, 1915; Paris. pp. 725-752.   7.

Benjamin, Dick, and J. J. Gravat; *Airborne Mine Sweeping*; Naval Aviation News, Aug. 1971. pp. 9-13.   8.

Benjamin, P.; *Battleship Mines and Torpedoes*; Revue of Revues, Vol. XXX, No. 1, Jul. 1904; New York. pp. 65-71.   9.

Bent, E.D.; *Floating Cable Helped to Destroy Menace of German Magnetic Mines*; Electrical News, Vol. LIV, No. 17, Sept. 1, 1945; Toronto. pp. 1-9.   10.

Benton, Elbert J.; *International Law and Diplomacy of the Spanish-American War*; Johns Hopkins Press, 1908; Baltimore. pg. 139.   11.

Berg, Ernst; *Die Seeminen im Krieg*; Dissertation, L. Gohring & Co., 1910; Augsburg.   12.

Bericht Über die Welt Ausstellung zu Paris; Wilhelm Braunmuller, 1867; Vienna.   13.

Bernay, Henri; *Defense Against Submarine Mines*; Translated by J. A. Mock in the Professional Notes, Journal of the U.S. Artillery, Vol. XXXI, No. 2, Mar-Apr. 1909; Ft. Monroe, Virginia. pp. 192-194.   14.

Bernotti, R.; *Il Potere Marittimo nella Grande Guerra*; Leghorn, 1920.   15.

1. Bertin, E.; *Le Transport des Mines par les Courants sous l'Action de la Houle;* Le Génie Civil, Vol. LXVI, No. 9, 1915; Paris. pg. 138.

2. Bethell, John; *On Firing Blasts Under Water by Galvanism;* Minutes of the Proceedings of the Institution of Civil Engineers. Abstracts of Papers and of the Conversations for the Session of 1838. Meeting of Apr. 24, 1838. pg. 35.

3. *Beskyttelse for Skibe mod forankrede Miner;* Tidsskrift for Sovaesen, Vol. XC, 1919; Copenhagen. pg. 220.

4. Betts, J. A.; *Torpedoes and Torpedo Warfare;* Telegraphic and Electrical Review, Vol. IV, No. 90, Nov. 1, 1876; London. pg. 282.

5. Betts, J. A.; *Torpedoes and Torpedo Warfare;* Telegraphic and Electrical Review, Vol. IV, No. 92, Dec. 1, 1876; London. pp. 304-305.

6. Bird, A.; *Submarine Mines, Their Use in Offensive Warfare;* Worlds Work, Vol. XXV, Dec. 1914; London. pp. 65-69.

7. Bishop, Farnhorn; *The Story of the Submarine;* The Century Company, 1916; London.

8. Bizot, M. (Translator); *Evolution de la Mine Sous-Marine;* by M. Sueter, Revue Maritime et Coloniale, Vol. CLXXX, 1890; Paris. pp. 267-329.

9. Blackman, Raymond V. B. (Editor); *Jane's Fighting Ships 1972-73;* Jane's Yearbooks, 1973; London.

10. Blakeslee, H. W.; *The Minelayer "Terror";* U.S. Naval Institute Proceedings, Vol. LXXXVI, No. 1, Jan. 1960; Annapolis, Maryland. p. 112.

11. Blic, Lt de vaisseau de; *Le Rôle de la Marine dans la Préparation et l'Exécution de la Campagne du James en 1862;* École de Guerre Navale au 1924. 71 p., 16 p. annexe. [Dactylographie.]

12. "Bloodhound"; Foreign Items; Army & Navy Journal, Vol. XVII, No. 6, Sept. 13, 1879; New York. pg. 110.

13. Bock, Ingeborg; *Die Entwicklung des Minenrechts von 1900 bis 1960;* (The Development of Legal Questions of Mines 1900-1960), 1963; Hamburg.

14. Boggs, H. Glenn; *A Study of the Confederate Naval Mining Capability in the U.S. Civil War;* Naval School, Mine Warfare Staff Officers Course Term Paper, May 11, 1970; Charleston, South Carolina.

Boling, Gerald; *The White Hat Skipper*; Navy, Vol. V, No. 8, Aug. 1962;   1.
    Washington, D.C. pp. 21-22.

Bolton, Reginald; *The Submarine Torpedo*; Engineer, Vol. LXXV, Mar. 10,   2.
    1893; London. pg. 208.

Bomber Command's Mine-Laying Successes: Planning and Carrying Out;   3.
    Illustrated London News, Vol. CCIV, Apr. 22, 1944; London. pg. 468.

Bone, M.; *Mining Cabinet;* Illustrated London News, Vol. CIC, Nov. 1,   4.
    1941; London. pp. 560-561.

Bonny, A. D.; *Effect of Non Contact Explosions on Warships Machinery*   5.
    Design; Engineering, Vol. CLXXV, Apr. 9, 1948; London. pp. 352-353
    and 356-357.

Bonny, A. D.; *Effect of Non Contact Explosions on Warships Machinery*   6.
    Design; Engineering, Vol. CLV, May 14, 1948; London. pp. 501-503.

Bonny, A. D.; *Effect of Non Contact Explosions on Warships Machinery*   7.
    Design; Engineering, Vol. CLV, May 28, 1948; London. pp. 525-526.

Bonny, A. D.; *Effect of Non Contact Explosions on Warship Machinery*   8.
    Design; Transactions of the Institute of Naval Architects, Paper
    No. 4, 1948; London.

Bourgois, Simeon; *Les Torpilleurs la Guerre Navale, et la Défense des*   9.
    Côtes; Librairie de la Nouvelle Revue, 1888; Paris.

Bowen, Frank C.; *Freak Ideas in Naval Warfare;* The Navy, Vol. XLIX,   10.
    No. 11, Nov. 1944; London. pg. 309.

Bowen, Frank C.; *Minelaying;* The Navy, Vol. XXXI, No. 4, Apr. 1926;   11.
    London. pp. 99-100.

Bowers, John V.; *Mine Warfare Channel Markers;* U.S. Naval Institute   12.
    Proceedings, Vol. LXXXIX, No. 9, Sept. 1863; Annapolis, Maryland.
    pp. 132-134.

Bowler, Roland Tomlin E.; *Mine Warfare;* U.S. Naval Institute Proceedings,   13.
    Vol. XC, No. 4, Apr. 1964; Annapolis, Maryland. pg. 4.

Bowler, Roland Tomlin E.; *Mine Warfare;* U.S. Naval Institute Proceedings,   14.
    Vol. XCI, No. 9, Sept. 1965; Annapolis, Maryland. pg. 4.

1. Bowler, Roland Tomlin E.; *Mine Warfare*; U.S. Naval Institute Proceedings, Vol. XCII, No. 3, Mar. 1966; Annapolis, Maryland. pp. 4-6.

2. Boynton, Charles B.; *The History of the Navy During the Rebellion*; D. Appleton, 2 volumes, 1867; New York.

3. Bradsher, Henry S.; *Mine Clearing Is Delayed For A Definition*; Sunday Star, (Evening Star), Vol. CXXI, No. 70, Sunday, Mar. 11, 1973; Washington, D.C. pg. A-10.

4. Bragadin, Marc' Antonio; *The Italian Navy in World War II*; U.S. Naval Institute, 1957; Annapolis, Maryland.

5. Brainard, Alfred P.; *Russian Mines on the Danube*; U.S. Naval Institute Proceedings, Vol. XCI, No. 7, Jul. 1965; Annapolis, Maryland. pp. 51-56.

6. Bramah, F., Jr.; *Account of Firing of Gunpowder Underwater, by the Voltaic Battery at Chatham, March 16, 1839, Under the Direction of Col. Pasley*; Minutes of the Proceedings of the Institution of Civil Engineers, Vol. V, Meeting of Mar. 19, 1839, Session 1839; pg. 50.

7. Brasseur, Pierre; *Le Guerre de Mines*; Nizet et Bastard, 1939; Paris.

8. Brassey; *The Naval Annual 1912*; Wm. Clowes & Son, London. pp. 313-316.

9. Bravetta, E.; *La Grande Guerra sul Mare*; Milan, 1925.

10. Breyer, Siegfried; *Soviet Minelayers and Minesweepers*; Soldat und Technik, Mar. 1968; Berlin. pp. 116-122.

11. *A Brief Description of Submarine Mines*; U.S. Navy Office of Naval Intelligence, U.S. Government Printing Office, 1917; Washington, D.C.

12. Bristol, Jack A.; *Here Come the Jap Mines*; Saturday Evening Post, Vol. CCXX, No. 38, Mar. 20, 1945; Philadelphia, Pennsylvania. pg. 12.

13. *British BB's Sunk by Mines*; Time, Vol. XXXVII, No. 17, Apr. 28, 1941; New York. pg. 26.

14. *British Coastal Minesweepers*; British Motor Ship, Vol. XXXV, No. 416, Nov. 1954; London. pp. 318-322.

15. *British Fishermen*; Time, Vol. XXXV, No. 9, Feb. 26, 1940; New York. pg. 37.

16. *British Leap to Meet Challenge of Nazi Onslaught in the North*; Newsweek, Vol. XV, No. 17, Apr. 22, 1940; New York. pp. 15-17.

British Mine Laying; United Service Gazette, Vol. CLXIII, Nov. 12, 1914; London. pp. 351-352.   1.

British Mine Locators; Electrician, Vol. CXXXVIII, Apr. 25, 1947; London. pg. 1080.   2.

British Strike with Reprisals to Meet the Challenge at Sea; Newsweek, Vol. XIV, No. 23, Dec. 4, 1939; New York. pp. 20-22.   3.

Brodie, Bernard; *A Guide to Naval Strategy;* Princeton Press, 4th Ed., 1958.   4.

Brodie, Bernard; *A Layman's Guide to Naval Strategy;* Princeton University Press, 1942.   5.

Brodie, Bernard; *Major Inventions and Their Consequences on International Politics, 1814-1918;* 1940; Chicago, Illinois.   6.

Brodie, Bernard; *Sea Power in the Machine Age;* Princeton University Press, 1941.   7.

Brookes, Ewart; *Glory Passed Them By;* Jarrolds, 1958; London.   8.

Brookfield, S. J.; *Mines and Counter-Measures;* Discovery, Jan. 1946; Cambridge. pp. 21-29.   9.

Brou, Willy Charles; *Les Mines Sous-Marines;* l'Armée, la Nation, 1949; Bruxelles. 6 p. [Croquis.]   10.

Brown, William Baker; *History of Submarine Mining in the British Army;* W. J. Mackay & Co., 1910; Chatham, Kent.   11.

La Bruyere, R.; *La Guerre de Mines;* Revue des Deux Mondes, Vol. LV, Feb. 15, 1940; Paris. pp. 726-738.   12.

Bryant, Samuel W.; *The Sea and the States;* Thomas Y. Crowell Company, 1947; New York.   13.

Buchard, H.; *Torpilles et Torpilleurs des Nations Étrangeres;* Berger-Levrault & Cie., 1889; Paris.   14.

Buckey, Mervyn Chandos; *Manual for the Instruction of Gunners of Mine Companies;* U.S. Coast Artillery, 1908; Ft. Worden, Washington.   15.

Bucknill, John Townsend; *The Destruction of the United States Battleship Maine;* U.S. Naval Institute Proceedings, Vol. XXIV, No. 3, Sept. 1898; Annapolis, Maryland. pp. 510-514.   16.

1. Bucknill, John Townsend; *On the Defences of Our Harbours by Submarine Mines*; Journal of the Royal United Service Institution, Vol. XXXI, Lecture XXI, Whole No. 139, 1887; London. pp. 263-296.

2. Bucknill, John Townsend; *Submarine Mines and Torpedoes*; Wiley, 1889; New York. xii, 225 p.

3. Bucknill, John Towsand; *Submarine Mines and Torpedoes (as Applied to Harbour Defence)*; John Wiley and Sons, Reprinted and revised from Engineering, 1889; New York.

4. Bucknill, John Townsend; *Torpedoes Versus Heavy Artillery: For Private Circulation*; 1872; Halifax, Nova Scotia.

5. *Building the Navy's Bases in World War II*; U.S. Government Printing Office, Vols. I and II, 1947; Washington, D.C.

6. *Building Wooden Ships for Iron Men*; Nautical Gazette, Vol. CXLVI, No. 10, Oct. 1952; New York. pp. 28-29, 49.

7. *Building Wooden Ships for Iron Men*; Marine Engineering Shipping Review, Vol. LVIII, No. 4, Apr. 1953; New York. pp. 50-54.

8. *Built to be Obliterated*; Armed Forces Journal, Vol. CVII, No. 26, Mar. 14, 1970; Washington, D.C. pg. 24.

9. Bultman, H. F. E.; *The Army Mine Planter Service*; Coast Artillery Journal, Vol. LXX, No. 6, Jun. 1919; Ft. Monroe, Virginia. pp. 469-472.

10. Bunker, Paul D.; *Forms for Records of Mine Planting*; Journal of the U.S. Artillery, Vol. LXIII, No. 3, May-Jun. 1915; Ft. Monroe, Virginia. pp. 331-335.

11. Bunker, Paul D.; *The Mine Defense of Harbors: Its History, Principles, Relation to the Other Elements of Defense, and Tactical Employment*; Journal of the U.S. Artillery, Vol. XLI, No. 2, Mar-Apr. 1914; Ft. Monroe, Virginia. pp. 129-170.

12. Bunker, Paul D.; *Submarine Mines*; Engineering Magazine, Vol. XLVII, Jun. 1914; New York. pp. 417-419.

13. *Buoyant Cables*; British Motor Ship, Vol. XXVI, No. 306, Jul. 1945; London. pg. 138.

14. *Buoyant Cables*; Electrical Review, Vol. CXXXVI, Jun. 8, 1945; London. pp. 829-832.

Buoyant Cables; Electrician, Vol. CXXXIV, Jun. 1, 1945; London.   1.
     pp. 487-489.

Burgoyne, A. H.; *Submarine Navigation - Past and Present;* E. P. Dutton,   2.
     1903; New York.

Busch, Fritz Otto; *Minen und Menschen von Peter Cornelissen;* W. Bischoff,   3.
     1933; Berlin.

(Bushnell, David); *Dictionary of American Biography;* Scribner's, Vol. II,   4.
     1958; New York.

Bushnell, David; (In the National Cyclopedia of American Biography),   5.
     Vol. IX, 1898; New York. pg. 244.

Bushnell, David; The Transactions of the American Philosophical Society   6.
     (held at Philadelphia for promoting useful knowledge). Transaction
     No. XXXVII, Vol. IV, published by Thomas Dobson, 1799; Philadelphia.
     pp. 303-312.

Butterworth, A.; *Development and Use of Magnetic Apparatus for Bomb and*   7.
     Mine Location; Journal of Institution of Electrical Engineers,
     Vol. XCV, Part 2, No. 48, Dec. 1948; London. pp. 645-652.

Cables That Float; Transport World, Vol. XCVIII, No. 3126, Jul. 12, 1945;   8.
     London. pp. 36, 39.

Cagle, Malcolm W. and Frank A. Manson; *Battle Report the War in Korea;*   9.
     Rinehart and Co., Inc., Vol. VI, 1952; New York.

Cagle, Malcolm W. and Frank A. Manson; *Wonsan: The Battle of the Mines;*   10.
     U.S. Naval Institute Proceedings, Vol. LXXXIII, No. 6, Jun. 1957;
     Annapolis, Maryland. pp. 598-611.

Callender, M. L.; *Floating Marine Torpedoes;* American Artisan, Vol. XII,   11.
     No. 7, Feb. 15, 1871; Chicago. pp. 99-100.

Callender, M. L.; *Floating Marine Torpedoes;* Van Nostrand's Engineering   12.
     Magazine, Vol. IV, No. 26, Feb. 1871; New York. pp. 237-240.

Callwell, C. E.; *The Dardanelles;* Houghton and Mifflin Company, 1919;   13.
     Boston and New York.

Calme, Barney; *The Raising of the U.S.S. CAIRO;* All Hands, No. 582,   14.
     Jul. 1965; Washington, D.C. pp. 25-27.

1. Calme, Byron E.; *Deadly and Sophisticated Navy's Mines Can Do Everything but Smell*; Wall Street Journal, Vol. CLXXIX, No. 92, May 10, 1972; New York. Front page and pg. 12.

2. Campbell, John; *Naval History of Great Britain*; A. Baldwin and Co., 8 volumes, 1818; London.

3. *Canadian Mine Sweepers*; Diesel Progress, Vol. XX, No. 9, Sept. 1954; New York. pp. 48-49.

4. Canfield, Eugene B.; *Notes on Naval Ordnance of the American Civil War, 1861-1865*; American Ordnance Association, 1960; Washington, D.C.

5. Capehart, E. E.; *The Mine Defense of Santiago Harbor*; U.S. Naval Institute Proceedings, Vol. XXIV, No. 4, Whole No. 88, Dec. 1898; Annapolis, Maryland. pp. 585-604.

6. *Captured German Mine-Laying Submarine*; Scientific American, Vol. CXV, No. 8, Aug. 19, 1916; New York. pg. 177.

7. Carter, W. R. and E. E. Duvall; *Ships Salvage and the Sinews of War*; U.S. Government Printing Office, 1954; Washington, D.C.

8. Cassell; *History of the Russo-Japanese War*; Cassell and Company, Ltd., Vol. I, 1904; London, Paris, New York and Melbourne.

9. Catlin, George L.; *Paravanes*; U.S. Naval Institute Proceedings, Vol. XLV, No. 7, Jul. 1919; Annapolis, Maryland. pp. 1134-1157.

10. *Centennial Exhibition - U.S. Department - Torpedoes*; Scientific American Supplement, Vol. II, No. 31, Jul. 29, 1876; New York. Front page and pg. 479.

11. Chadwick, French Ensor; *Russo-Japanese War 1903-05*; Collection of Notes; Compiled by Navy Department Library.

12. Champlin, G. F.; *Piasecki Develops New Minesweeping Technique*; American Helicopter, Vol. XXXVIII, Mar. 12, 1955; New York. pg. 6.

13. *A Chapter on Submarine Apparatus*; Colburn's United Service Magazine, Vol. CXV, Hurst and Blackett, Successors to Henry Colburn, 1868; London. pp. 217-223.

14. Chardonneau, F.; *Des Torpilles Russes sur le Danube à Soulina et à Batoum*; Revue Maritimeet Coloniale, Vol. LVII, 1878; Paris. pp. 156-177.

Chardonneau, F.; *Des Torpilles Russes sur le Danube à Soulina et à Batoum*; Revue Maritimeet Coloniale, Vol. LVIII, 1878; Paris. pp. 75-101. 1.

Chardonneau, F.; *Of the Russian Torpedoes on the Danube at Soulina and Batum*; Journal of the Royal United Service Institution, Vol. XXII, No. 96, 1879; London. pp. 735-745. 2.

Chardonneau, F.; *Of the Russian Torpedoes on the Danube at Soulina and Batum*; Journal of the Royal United Service Institution, Vol. XXII, No. 97, 1879; London. pp. 1049-1065. 3.

H.M.S. Charges; Time, Vol. XXXV, No. 4, Feb. 19, 1940; New York. pg. 23. 4.

Chart of Submarine Sinkings; February 1917 to April 1918; Engineering, Vol. CLIV, No. 3998, Aug. 28, 1942; London. pp. 175-176. 5.

La Chasse Aux Mines; Prépare pour l'Amerauté par le Ministère de l'Information, Her Majesty's Stationery Office, 1943; Londres. 185 p. 6.

La Chasse Aux Mines; Prépare pour l'Amerauté par le Ministère de l'Information, Her Majesty's Stationery Office, 1947; Londres. 185 p. 7.

"Chasseur"; *A Study of the Russo-Japanese War I. The Naval Campaign*; Blackwood's Magazine, Vol. CLXXVII, No. 571, Jan. 1905; London. pp. 144-174. 8.

Chatterton, E. Keble; *Dardanelles Dilemma*; Rich and Cowan, 1935; London. 9.

Chatterton, Howard A.; *The Minesweeping/Fishing Vessel*; Professional Notes, U.S. Naval Institute Proceedings, Vol. XCVI, No. 6, Jun. 1970; Annapolis, Maryland. pp. 121-125. 10.

A Chronology of U.S. Involvement; National Observer, Vol. XII, No. 5, Feb. 3, 1973; Silver Spring, Maryland. pg. 17. 11.

Chuckles Mixed With Sneezes in Mine Story; Science News Letter, Vol. LXI, Apr. 19, 1952; Washington, D.C. pg. 249. 12.

Churchill, W. S.; *The World Crisis*; Charles Scribners Sons, 1931; New York. 13.

Circuit Closers; Engineering, Vol. XXI, Feb. 4, 1876; London. pp. 96-98. 14.

Circuit Closers; Engineering, Vol. XXI, Feb. 25, 1876; London. pg. 153. 15.

1. *Circuit Closers;* Engineering, Vol. XXI, Mar. 10, 1876; London. pp. 184-185.

2. *Circuit Closers;* Engineering, Vol. XXI, Mar. 24, 1876; London. pp. 224-225.

3. *Circuit Closers;* Engineering, Vol. XXI, May 19, 1876; London. pp. 404-406.

4. *Circuit Closers;* Engineering, Vol. XXI, Jun. 9, 1876; London. pp. 475-577.

5. *Circuit Closers;* Engineering, Vol. XXI, Jun. 30, 1876; London. pp. 549-550.

6. *Civil War Naval Chronology 1861-1865;* Naval History Division, Office CNO Navy Department. U.S. Government Printing Office, 6 volumes with index, 1966; Washington, D.C.

7. *Civil War Naval Chronology, Parts I, II, III, IV, VI.* U.S. Government Printing Office, 1961, 1962, 1963, 1964.

8. Clark, Frank S.; *Mine Defense Today and Tomorrow;* Coast Artillery Journal, Vol. LXXI, No. 3, Sept. 1929; Ft. Monroe, Virginia. pp. 181-198.

9. Clark, Thomas; *Naval History of the United States;* M. Carey, Vol. I, Jan. 3, 1814; Philadelphia. pp. 63-74.

10. Clark, Thomas; *Sketches of the Naval History of the United States;* 1813; Philadelphia.

11. Clark, William Bell (Editor); *Naval Documents of the American Revolution;* Navy Department, Vol. I - American Theatre, Dec. 1, 1774 - Sept. 2, 1775; European Theatre, Dec. 6, 1774 - Aug. 9, 1775; 1964; Washington, D.C. pp. 1088-1089, 1244, 1274.

12. Clark, William Bell (Editor); *Naval Documents of the American Revolution;* Navy Department, Vol. II, 1966; Washington, D.C. pp. 953-956, 1099-1100, 1324-1325, 1050, 1084.

13. Clark, William Bell (Editor); *Naval Documents of the American Revolution;* Navy Department, Vol. III, 1968; Washington, D.C.

14. Clark, William Bell; *When the U-Boats Came to America;* Little, Brown and Co., 1929; Boston.

15. Clarke, G. S.; *Submarine Mines in Relation to War;* Royal Artillery Institute, 1890; Woolwich.

Clarke, G. S.; *The War and Its Lessons;* U.S. Naval Institute Proceedings,   1.
    Vol. XXVI, No. 1, Mar. 1900; Annapolis, Maryland. pp. 127-141.

Clarke, Lt-Col. Sir Georges S. and James R. Thursfield; *The Navy and the*   2.
    Nation or Naval Warfare and Imperial Defense; Murray, 1897; London.
    344 p.

"Claudeville"; *Importance des Mines Sous-Marines dans les Guerres Maritime;*   3.
    Revue Maritime Nouvelle Serie, Vol. I, 1926; Paris. pp. 305-323
    and 627-646.

Clearing the Mines; Time, Vol. CI, No. 7, Feb. 12, 1973; Chicago and   4.
    New York. pp. 23-24.

Cleator, P. E.; *Weapons of War;* Crowell, 1968; New York.   5.

Closing the North Sea with Mines Was the Job of the Gob; Literary Digest,   6.
    Vol. LX, Feb. 15, 1919; New York. pp. 94-98.

Clouet, Alain; *Notes Historiques sur la Guerre des Mines;* Groupe d'Études   7.
    de la Guerre de Mines, impr. D.C.A.N., 1960; Brest. 133 p.

Clowes, William Laird; *The Royal Navy: A History From the Earliest Times*   8.
    to the Present; Clowes, Ltd., 7 volumes, 1897-1903.

Clowes, William Laird; *The Royal Navy: A History From the Earliest Times*   9.
    to the Present; Sampson Low, Marston and Co., Vol. III, 1898;
    London. pp. 383, 406.

Cluverius, Wat Tyler; *Planting a War Garden;* U.S. Naval Institute Pro-   10.
    ceedings, Vol. XLV, No. 3, Mar. 1919; Annapolis, Maryland. pp. 333-338.

Coast Defense; Journal of the United States Artillery, Vol. XXII, No. 3,   11.
    Nov-Dec. 1904; Ft. Monroe, Virginia. pp. 215-265.

Coast Defense by Submarine Mines; Scientific American, Vol. CVI, No. 20,   12.
    May 18, 1912; New York. pg. 444.

Coastal Command; Great Britain Air Ministry, Macmillan Co., 1943; New York.   13.

Colden, Cadwalder D.; *The Life of Robert Fulton (1817);* Reviewed in the   14.
    Quarterly Review, John Murray, Vol. XIX, Apr. and Dec. 1818; London.

Coleman, Frank; *Underwater Detection by Helicopter;* Journal of the American   15.
    Helicopter Society, Vol. XLVI, May 1957; New York. pg. 12.

1. Coletta, Paola E.; *Naval Mine Warfare*; Navy, Vol. II, No. 11, Nov. 1959; Washington, D.C. pp. 16-24.

2. Coletta, Paola E.; *Naval Mine Warfare*: U.S. Naval Institute Proceedings, Vol. LXXXV, No. 11, Nov. 1959; Annapolis, Maryland. pp. 82-126.

3. Colombos, John Constantine; *The International Law of the Sea*; Longmans, 1959; London.

4. Colombos, John Constantine; *The International Law of the Sea*; Longmans, 1967; London. pg. 533.

5. Comey, Robert W.; *The Mine Forces: Problems of Advanced Technology*; U.S. Naval Schools, Mine Warfare, Nov. 1960; Charleston, South Carolina.

6. *Commerce Protection and Submarine Mines*; Army Navy Gazette, Vol. XLVII, 1960; London. pp. 276-277.

7. Committee on Floating Obstructions and Submarine Explosive Machines; *Report on Active Obstructions for the Defence of Harbours and Channels etc.*; Her Majesty's Stationery Office, 1868; London.

8. Condon, J. F. and H. M. Koslow; *Magnetic Mine Sweeping System*; Electrical Engineering, Vol. LXVII, No. 9520, Dec. 1948; New York. pp. 1196-1197.

9. *Conférences Faites à Bord du Vaisseau École le "Louis XIV" sur l'Emploi des Torpilles pour la Défense des Ports.* 235 p. et 11 planches.

10. Congress of the Institute of International Law, Army and Navy Journal, Vol. XLIV, No. 5, Sept. 29, 1906; New York. pg. 115.

11. *HMS CONISTON*; Joint Services Recognition Journal, Vol. IX, Jun. 1954; London. pp. 152-153.

12. Conness, Leland; *Submarine Mines*; Munsey's Magazine, Vol. LV, No. 2, Jul. 1915; New York. pp. 257-265.

13. *Construction Plans (Navy) Minesweepers*; Military Review, Vol. XXXVI, No. 6, Sept. 1956; Ft. Leavenworth. pg. 69.

14. Conti, A.; *Bateaux de Pêche en Travail de Guerre; Détermination d'un Chenal de Sécurité*; l'Illustration, Vol. CCIV, No. 5051, Dec. 23, 1939; Paris. pp. 457-460.

15. *Controlled Submarine Mines World War II Operational Record 1941-1945*; U.S. Submarine Mine Depot, 1949; Ft. Monroe, Virginia.

Copter-Towed Sea Sled Replaces Minesweepers; Product Engineering, Vol. XLII,   1.
   Apr. 12, 1971; New York. pp. 13-14.

Cook, Gilbert; *A Recording Submarine Depth Meter;* Engineering, Vol. CIX,   2.
   No. 2828, Mar. 1920; London. pg. 333.

Cooper, James Fenimore; *The History of the United States Navy;* Lea and   3.
   Blanchard, 2 volumes, 1839; Philadelphia.

Cooper, James Fenimore; *History of the Navy of the United States of*   4.
   America; Oakley & Mason, 1866; New York.

Corbett, Julian S.; *Some Principles of Maritime Strategy;* Longmans, Green,   5.
   1911; London.

Corbin, Diana F. M.; *A Life of Mathew Fontaine Maury;* 1888; London.   6.

Counter Mines; U.S. Navy, U.S. Government Printing Office, 1905;   7.
   Washington, D.C.

Countering the Magnetic Mine; Engineer, Vol. CLXIX, Mar. 15, 1940; London.   8.

Countering the Magnetic Mine; Nature, Vol. CXLV, Mar. 16, 1940; London.   9.
   pg. 415.

Cowie, John S.; *British Mines and the Channel Dash;* U.S. Naval Institute   10.
   Proceedings, Vol. LXXXIV, No. 4, Apr. 1958; Annapolis, Maryland.
   pp. 39-47.

Cowie, John S.; *The British Sea Campaign 1939-1945;* Journal of the Royal   11.
   United Service Institution, Vol. XCIII, Feb. 1948; London. pp. 22-39.

Cowie, John S.; *The Mine as a Tool of Limited War, Comments and Discussions;*   12.
   U.S. Naval Institute Proceedings, Vol. XCIII, No. 10, Oct. 1967;
   Annapolis, Maryland. pg. 106.

Cowie, John S.; *Mines, Minelayers and Minelaying;* Oxford University Press,   13.
   1949; London and New York.

Cowie, John S.; *Mines Were Laid;* The Navy, Vol. L, No. 5, May 1945; London.   14.
   pp. 140-145.

Cowie, John S.; *Minelayers;* Journal of the Royal United Service Institution,   15.
   Vol. C, Nov. 1955; London. pp. 602-610.

1. Cowie, John S.; *Minelaying From Aircraft;* Journal of the Royal United
     Service Institution, Vol. LXXXIX, May 1944; London. pp. 137-139.

2. Cowie, John S.; *Minelaying in the War;* Journal of the Royal United Service
     Institution, Vol. XCI, Aug. 1946; London. pg. 413.

3. Cowie, John S.; *Mining of Inland Waters;* Journal of the Royal United
     Service Institution, Vol. CV, Nov. 1960; London. pp. 532-534.

4. Cowie, John S.; *The Potential Menace of the Sea Mine;* Brassey's Annual,
     No. 62, 1952; London. pp. 194-200.

5. Cowie, John S.; *Role of the United States Navy in Mine Warfare;* U.S.
     Naval Institute Proceedings, Vol. XCI, No. 5, May 1965; Annapolis,
     Maryland. pp. 52-63.

6. Cowie, John S.; *Russian Mines on the Danube;* U.S. Naval Institute Pro-
     ceedings, Vol. XCII, No. 11, Nov. 1966; Annapolis, Maryland.
     pp. 133-134.

7. Cowie, John S.; *Sea Mines;* The Navy, Vol. LII, No. 8, Aug. 1948; London.
     pp. 280-281.

8. Craighill, W. E.; *Foreign Systems of Torpedoes as Compared With Our Own;*
     United States Engineer School, Paper No. XII, 1888; Willets Point,
     New York Harbor.

9. Cranford, L. Cope; *The Paravane Adventure;* Hodder and Stoughton, 1919;
     London.

10. Craven, Wesley Frank and James Lea Cate; *The Army Air Forces in World
     War II;* Vol. V, The Pacific: Matterhorn to Nagasaki Jun. 1944 to
     Aug. 1945, University of Chicago Press, 1953.

11. *Credit to Mine Sweepers;* Letters to the Editors, All Hands, No. 330,
     Sept. 1944; Washington, D.C. pp. 36, 59.

12. *Crouching Dragons;* Time, Vol. XLVI, No. 27, Dec. 31, 1945; New York. pg. 26.

13. Crouse, George M.; *Submarine Mines;* Yale Scientific Monthly, Jun. 1898;
     New Haven, Connecticut.

14. Crowley, R. O.; *Confederate Torpedo Service;* Century Magazine, Vol. LVI,
     No. 2, Jul. 1898; New York. pp. 290-300.

Crowther, James Gerald and R. Wheddington; *Science at War;* Her Majesty's   1.
    Stationery Office, 1947; London. 185 p.

Crowther, James Gerald and R. Wheddington; *Science at War;* Philosophical   2.
    Library, 1948; New York.

Cruiser-Minelayers; Professional Notes, U.S. Naval Institute Proceedings,   3.
    Vol. XLIX, No. 4, Apr. 1923; Annapolis, Maryland. pp. 675-676.

Cummings, Joseph D.; *Greater Mine Warfare Capability Urged for U.S. Navy;*   4.
    Navy, Vol. X, No. 5, May 1967; Washington, D.C. pg. 6.

Cummings, Joseph D.; *Whatever Became of the Mine;* U.S. Naval Institute   5.
    Proceedings, Vol. XCII, No. 6, Jun. 1966; Annapolis, Maryland.
    pp. 117-118.

Cunningham, Andrew B.; *A Sailor's Odyssey;* Hutchinson and Co., 1951;   6.
    Dutton, New York.

Cutbush, James; *A System of Pyrotechny;* Clara F. Cutbush, 1825;   7.
    Philadelphia.

Dahl, E. M.; *On Minesagning og Minerydning;* Tidsskrift for Sovaesen,   8.
    Vol. C, 1929; Copenhagen. pp. 434, 481, 534.

Dallin, David J.; *The Real Sovet Russia;* Yale University Press, 1947.   9.

Daly, Robert W.; *Soviet Navy;* Edited by M. G. Saunders; 1958; Praeger,   10.
    New York.

Danger Awash; Newsweek, Vol. XXVII, No. 10, Mar. 11, 1946; Washington,   11.
    D.C. pp. 39-40.

Dangerous Task in London's Docks; Illustrated London News, Vol. CCXXX,   12.
    Feb. 2, 1957; London. pg. 171.

Daniel, C. S.; *Degaussing;* Journal of the Institution of Electrical   13.
    Engineers, Vol. XCIII, Part 1, Jun. 21, 1946; London. pp. 247-248.

Daniels, Josephus; *The Northern Barrage and Other Mining Activities;*   14.
    Publication No. 2, U.S. Government Printing Office, 1920;
    Washington, D.C. 146 p.

Danish Naval Plans; United Service Magazine, Vol. XL, No. 975, Feb. 1910;   15.
    London. pg. 553.

Danske Mineovelser i Øresund; Tidsskrift for Sovaesen, Vol. LXVI, 1895;   16.
    Copenhagen. pg. 391.

1. *DAPAC (Danger Areas in the Pacific)*; U.S. Hydrographic Office, 2nd Ed., 1960; 3rd Edition, 1964; U.S. Government Printing Office, Washington, D.C.

2. *Dardanelles Campaign - World War I*; Encyclopedia Britannica, Vol. XXIII, 1963 Edition; Chicago. pp. 782-784.

3. Dauch, Lt de vaisseau; *Pose de Mines par Navires de Surface Guerre 1914-1918*; École de Guerre Navale Année 1930; Exempl No. 23, sl. sn. sd. 67 p.

4. Daudenart, Major d'État-Major L. G.; *La Guerre Sous-Marine et les Torpedoes*; C. Muquardt, J. Dumaine, 1872; Bruxelles-Paris. 93 p.

5. Davelny, Contre-Amiral; *Les Enseignements Maritimes de la Guerre. (Aout 1914-Novembre 1918)*; A. Challumel, 1919. 167 p.

6. Davelny, Rene; *Étude sur le Combat Naval*; Berger-Levrault & Cie., One Volume, 1902; Paris.

7. Davidson, Hunter; *Electrical Torpedoes*; Army and Navy Journal, Vol. XI, No. 52, 1874; New York. pg. 825.

8. Davidson, Hunter; *Torpedoes in Our War*; U.S. Naval Institute Proceedings, Vol. XXIV, No. 2, Jun. 1898; Annapolis, Maryland. pp. 349-354.

9. Davies, G.; *Submarine Shells or Torpedoes*; The Engineer, Vol. XIX, Mar. 10, 1865; London. pg. 159.

10. Davis, Ewing O.; *Hundred Years Ago and Today*; Military Engineer, Vol. LIV, No. 358, 1962; Washington, D.C. pp. 110-111.

11. Davis, G. H.; *Acoustic Mine: New Terror of the Under-Sea War*; Illustrated London News, Vol. CIC, Sept. 20, 1941; London. pg. 361.

12. Davis, G. H.; *Minesweepers and Minelayers: Key Vessels of Modern Naval Warfare*; Illustrated London News, Vol. CCXXIII, Aug. 1, 1958; London. pp. 182-183.

13. Davis, G. H.; *New Types of Mines Countered by the Navy*; Illustrated London News, Vol. CCVIII, Feb. 23, 1946; London. pp. 200-201.

14. Davis, J. H.; *Magnetic Mines and Degaussing of Ships*; Baltimore Engineer, Vol. XVI, No. 4, Oct. 1941; Baltimore. pp. 1-7.

Davis, Noel; *Removal of the North Sea Mine Barrage*; National Geographic,   1.
     Vol. XXXVII, Feb. 1910; Washington, D.C.

Davis, Noel (Editor); *Sweeping the North Sea Mine Barrage 1919*; North   2.
     Sea Mine Sweeping Detachment; Printed by J. D. McGuire, 1919;
     New York.

Davis, Richmond P.; *Development of the Submarine Mine in the United States*   3.
     Service; Journal of the U.S. Artillery, Vol. XXXIX, No. 1, Jan-Feb.
     1913; Ft. Monroe, Virginia. pp. 15-32.

Davis, Richmond P.; *Submarine Mines and Mining*; Journal of the American   4.
     Society of Naval Engineers, Vol. XX, 1908; Washington, D.C.
     pp. 674-691.

Davis, Richmond P.; *Submarine Mines and Mining*; Journal of the U.S.   5.
     Artillery, Vol. XXIX, No. 3, May-Jun. 1908; Ft. Monroe, Virginia.
     pp. 225-239.

Davy Jones' Sound Effects; Time, Vol. XLVII, No. 17, Apr. 29, 1946; New York.   6.
     pp. 90-91.

Dawson, W.; *Offensive Torpedo Warfare*; Journal of the Royal United Service   7.
     Institution, Vol. XV, No. 62, Jan. 1872; London. pp. 86-111.

Death in Donegal; Time, Vol. XLI, No. 21, May 24, 1943; New York. pg. 32.   8.

The Declaration of Paris; Supplement to the American Journal of Interna-   9.
     tional Law, Apr. 1907. pg. 89.

Defeating Magnetic Mines; South African Engineer and Electrical Review,   10.
     Vol. XXXV, No. 331, Nov. 1945; Johannesburg. pp. 29, 31, 39.

Defense Against Submarine Mines; Journal of the U.S. Artillery, Vol. XLIV,   11.
     Jul-Dec. 1915; Ft. Monroe, Virginia. pp. 229-230.

Defense Against Submarine Mines; Journal of the American Society of Naval   12.
     Engineers, Vol. XXVI, 1914; Washington, D.C. pp. 1400-1401.

Defense Against Submarine Mines; Mechanical World, Vol. LVI, 1914; London.   13.
     pg. 81.

Degaussing; Electrical Review, Vol. CXXXVIII, Apr. 12, 1946; London.   14.
     pp. 565-566.

Degaussing; Journal of the Institution of Electrical Engineers, Vol. XCIII,   15.
     Part 1, No. 70, 1946; London. pp. 429-455.

1. *Degaussing;* Journal of Institution of Electrical Engineers, Vol. XCIII, Part 1, No. 71, 1946; London. pp. 488-524.

2. *Degaussing the Queen Elizabeth;* Time, Vol. XXXV, No. 12, Mar. 18, 1940; New York. pg. 27.

3. De Gouy, Jean Baptiste and C. R. Mathieu; *Le Guerre Sous-Marine de 1917;* Payot and Co., 1918; Paris.

4. De Gouy, Jean Baptiste and C. R. Mathieu; *Pour en Finir Avec les Sous-Marines;* Payot and Co., 1918; Paris.

5. DeGreene, K. B.; *Systems Psychology;* McGraw-Hill, 1970; New York.

6. DeKerc'hoat, L.; *La Protection Interieure contre les Torpilles et les Mines;* Journal de Marine le Yacht, Vol. XXXVI, No. 1857, Oct. 11, 1913; Paris. pp. 653-654.

7. Delafield, R.; *Art of War in Europe in 1854, 1855 & 1856;* George W. Bowman, Printer, 1860; Washington, D.C.

8. Delegue, Cap de Frégate; *Conférence d'Information sur les Mines Sous-Marines;* Commission d'Études Pratiques Mines Grenades, 1926. 26 p.

9. Delpeuch, Lt de vaisseau Maurice; *La Première Tentative de Guerre Sous-Marine;* La Contemporaine, 1901; Paris. 12 p. (Extrait de *La Contemporaine,* Oct. 25, 1901, pp. 217-228.)

10. Demigny, A.; *La Faillite de la Marine. Étude Critique Maritime et Militaire;* Berger-Levrault & Cie., 1899; Paris.

11. *Demijohn Torpedoes;* Scientific American, Vol. VIII, No. 2, Jan. 10, 1863; New York. pg. 19.

12. Dempewolff, Richard F.; *Mother of the Mine Sweepers;* Popular Mechanics, Vol. XCVII, No. 2, Feb. 1952; Chicago. pp. 97-104.

13. Denlinger, Sutherland and Charles P. Gary; *War in the Pacific;* Robert M. McBride & Co., 1936; New York. pp. 110-111.

14. *Den Nye Ordning af det franske Søminervaesen;* Tidsskrift for Sovaesen, Vol. LXV, 1894; Copenhagen. pg. 43.

15. *Den undersoiske Mine ved Liaotung Halvøerunder den russisk-japanske Krig 1904-05;* Tidsskrift for Sovaesen, Vol. LXXXIII, 1912; Copenhagen. pg. 1.

Deployment Report: JUK Group Mediterranean and EASTLANT Operations (U); CTG 83.4/67.6/60.6/87.0; 15 February-14 June 1968; FB/CCL-20/002:lc 3120; Ser: 0013; Commander Carrier Division TWENTY: Jul. 3, 1968. CONFIDENTIAL.   1.

Derelict Mines a Peace Peril; Scientific American, Vol. CXX, No. 10, Mar. 1, 1919; New York. pg. 196.   2.

Derelict Mines of Japs Cause Problem on Coast; National Underwriter, Vol. LI, Dec. 4, 1947; Cincinnati, Ohio. pg. 14.   3.

Der Klein Kueuzer "Cassini" Mineleger; Schifbau, Vol. XIII, 1911; Berlin. pg. 21.   4.

De seneste Forwoy med undersoiske Miner; Tidsskrift for Sovaesen, Vol. XLII, 1871; Copenhagen. pg. 58.   5.

Destroying a Russian Mine; Scientific American, Vol. XCII, No. 20, May 20, 1905; New York. pg. 406.   6.

Destruction of a U.S. Steamer by a Torpedo; Scientific American, Vol. X, No. 11, Mar. 12, 1864; New York. pg. 165.   7.

Detection, Navigation and Electronic Equipment; International Defense Review, No. 4, 1968; Intucivia, Geneva. pp. 258-259.   8.

Detonation of Submarine Mines by Electricity; Scientific American Supplement, Vol. LXXXI, No. 2102, Apr. 15, 1916; New York. pg. 253.   9.

Detonation of Submarine Mines by Means of Sound Waves; Scientific American, Vol. XCIV, No. 8, Feb. 24, 1906; New York. pg. 170.   10.

The Development of the Science of Mine-Sweeping; Journal of the Royal United Service Institution, Vol. LX, No. 440, Nov. 1915; London. pp. 383-403.   11.

Devize (ou Deveze), A.; *Calcul d'un Projectile Sous-Marine Explosif;* G. Villers, 1866; Paris. 32 p. et 11 planches.   12.

Dibos, M.; *De la Recherche et du Dragagedes Torpilles Vigilontes;* Memorial Society, Civil Engineer of France, Oct. 1904; Paris.   13.

Dickinson, H. W.; *Robert Fulton, Engineer and Artist, His Life and Works;* John Lanes, 1913; London.   14.

Didelot, Carl; *La Défense des Côtes d'Europe. Étude Descriptive au Double Point de Vue Militarie et Maritime;* Berger-Levrault & Cie., 1895; Paris. 540 p.   15.

1. *Die durch Minen an den Russischen Port Arthur-Schiffen Verursachten Bechadingungen und die Reparaturen der Verletzter Schiffe;* Ueberall, Vol. VIII, No. 8, 1905; Berlin. pp. 1-2.

2. *Digest of London Naval Treaty of 1930;* U.S. Department of State Conference Series No. 4.

3. *Documentary History of the War (WW I);* The Times, Vol. XI, Naval Part 4, 1920; London. pp. 121-132.

4. Doflein, M.; *Der Kampf der Minensuch Flotillen;* Königliche Fredrich-Wilhelms Universität-Institut für Meerskunde, 1918; Berlin.

5. Dohna-Schlodien, Burgrave Nicolas zu; *La "Möwe", Ses Croisières et Ses Aventures;* Traduit de l'Allemande par René Jouan. Payot, 1929; Paris. 222 p.

6. Dollo, A.; *Les Mines et les Torpilles;* Revue Trimestrielle Canadienne, Vol. I, 1915; Montreal. pp. 55-56.

7. Dommett, William Erskine; *Submarine, Torpedoes, and Mines;* Whittaker and Co., 1915; New York.

8. Dommett, William Erskine; *Submarine Vessels;* Whittaker and Co., 1915; London.

9. Domville-Fife, Charles W.; *Submarines, Mines, and Torpedoes in the War;* Hodder and Stoughton, 1914; London.

10. Dooly, William G.; *Great Weapons of World War I;* Army Times Publication, 1969; Washington, D.C.

11. Dorling, H. Taprell ("Taffrail"); *The Navy in Action;* Hodder and Stoughton Ltd., 1940; London.

12. Dorling, H. Taprell ("Taffrail"); *Swept Channels;* Hodder and Stoughton Ltd., 1935; London.

13. (Dorthea); Naval Chronicle, Vol. XIV, Joyce Gold, 1805; London. pp. 341-342.

14. Doty; *On the Torpedoes Used by the Confederate States;* Army and Navy Journal, Vol. II, No. 9, Oct. 22, 1864; New York. pp. 138-139.

-29-

The Downfall of Spain, Naval History, 1898; Little Brown, 1900; Boston.   1.

Duke, Irving Terri; *Mine Warfare;* U.S. Naval Schools, Mine Warfare,   2.
  Charleston, South Carolina. (undated)

Dukers, Capitaine de Frégate; *La Guerre de Mines 1914-1918;*   3.
  Fascicules 1, 2, 3, 4. pg. 733.

H.M.S. Dunbar; Time, Vol. XXXV, No. 4, Jan. 22, 1940; New York. pg. 32.   4.

Duncan, Robert C.; *America's Use of Sea Mines;* U.S. Naval Ordnance   5.
  Laboratory, 1962; White Oak, Maryland.

Dunn, John M.; *Notes on Water Work in Submarine Mines;* Coast Artillery   6.
  Journal, Vol. LVII, No. 2, Aug. 1922; Ft. Monroe, Virginia. pp. 143-159.

Dunsany, Admiral Lord; *On the Laws and Customs of War as Limiting the*   7.
  Use of Fire-Ships Explosion Vessels, Torpedoes, and Submarine
  Mines; Journal of the Royal United Service Institution, Vol. XXII,
  No. 95, 1878; London. pp. 271-290.

Durassier, Edward; *Aide Memoire de l'Officer de Marine;* Henri Charles-   8.
  Lavauzelle, 1895, 1898 and 1903 Editions.

Dzienisiewicz, Cap de Corvette; *Emploi des Sous-Marines et des Mines*   9.
  dans la Baltique par les Allemands Pendant la Grande Guerre;
  École de Guerre Navale Session 1935-36, Exempl. No. 9.
  40 p. et annexes.

The Early History of Torpedoes; Journal of the Royal Society of Arts,   10.
  Vol. XXV, 1877; London. pp. 881-882.

Earp, G. Butler; *History of the Baltic Campaign of 1854;* Bentley, 1857;   11.
  London.

The Eastern War, Turkish Divers Removing Russian Torpedoes; Frank Leslie's   12.
  Illustrated Newspaper, Vol. XLIV, Aug. 18, 1877; New York.
  pp. 407-408.

"E.B."; *U. Baade til Mineudlaegning;* Tidsskrift for Sovaesen, Vol. LXXXIII,   13.
  1912; Copenhagen. pg. 434.

Edmonds, James E.; *Submarine Mining, The Life and Death of One of the*   14.
  Corps Offspring 1871-1905; Royal Engineers Journal, Vol. LXX,
  No. 4, Dec. 1956; Chatham Kent. pp. 382-386.

1. Edwards, Harry William; *A Naval Lesson of the Korean Conflict*; U.S. Naval
    Institute Proceedings, Vol. LXXX, No. 12, Dec. 1954; Annapolis,
    Maryland. pp. 1337-1340.

2. Edwards, Kenneth; *Mines Were Laid*; The Navy, Vol. LI, No. 6, Jun. 1946;
    London. pp. 199-202.

3. Edwards, Kenneth; *Mines Were Laid*; U.S. Naval Institute Proceedings, Vol.
    LXXIII, No. 1, Jan. 1947; Annapolis, Maryland. pp. 106-108.

4. *The Effect of Mines and Torpedoes on Merchant Ships*; U.S. Naval Institute
    Proceedings, Vol. XLIV, No. 5, May 1918; Annapolis, Maryland. pg. 1110.

5. Egan, Richard; *Hanoi's Troubles*; National Observer, Vol. XI, No. 25,
    Week ending Jun. 17, 1972; Silver Spring, Maryland. pg. 3.

6. von Ehrenkrook, Fredrich (Lt de vaisseau L.); *History of Submarine Mining
    and Torpedoes*; U.S. Engineer Professional Papers, Paper No. 1
    (translated by Sgt. Major Frederick Martin), Battalion Press, 1881;
    Willets Point, New York.

7. von Ehrenkrook, Fredrich (Lt de vaisseau L.); *Die Fisch Torpedoes*; Ihre
    Historische Entwickelung, Einrichtung, Verwendung und Bekamfung,
    sowie deren Einfluss auf zukunftige Seekriege. Mittler und Sohn,
    1878; Berlin. 92 p. und figures.

8. von Ehrenkrook, Fredrich (Lt de vaisseau L.); *Geschichte der Seeminen und
    Torpedos*; Mittler und Sohn, 1878; Berlin. 77 p.

9. Eisner, Will; *Naval Ordnance*; Sterling Publishing Co., 1960; New York.

10. *Electric Mine for Harbor Defense*; Illustrated American, Vol. XXIII,
    Mar. 26, 1898; New York. pp. 402-403.

11. *Electric Torpedoes*; Army and Navy Journal, Vol. XI, No. 36, Apr. 18, 1874;
    New York. pg. 574.

12. *The Electrical Torpedo*; Scientific American, Vol. XIII, No. 8, Aug. 19,
    1865; New York. pg. 113.

13. *Electrical Warfare*; Army and Navy Journal, Vol. XI, No. 38, May 2, 1874;
    New York. pg. 600.

14. *Electrically Exploding Torpedoes*; American Artisan, Vol. XII, No. 72,
    May 31, 1871. pg. 344.

Electro-Contact Torpedoes; Engineering, Vol. XXIII, Mar. 2, 1877; London.   1.
       pp. 157-158.

Electro-Mechanical Torpedoes; Engineering, Vol. XXIII, Mar. 16, 1877;   2.
       London. pp. 214-216.

Electro-Mechanical Torpedoes; Engineering, Vol. XXIII, Jun. 25, 1877;   3.
       London. pp. 455-456.

Elia   Giovanni Emanuelle; *Selbstatige, Ankervorrichtung für Seeminen mit*   4.
       Lot zur Regelung der Tauchtiefe; Zeitschrift für das gesamte Schiess
       und Sprengstoffwesen, Vol. VIII, 1913; Munich. pp. 236-237.

The Elia Mine; U.S. Naval Institute Proceedings, Vol. XXXVIII, No. 2,   5.
       Jun. 1912; Annapolis, Maryland. pg. 788.

The Elia System of Construction and Placing Submarine Mines; Le Génie   6.
       Civil, Vol. LXII, No. 10, Jan. 4, 1913; Paris. pp. 189, 192.

Eliot, G. F.; *How Russia Threatens Our Sea Power;* Colliers, Vol. CXXXII,   7.
       Sept. 4, 1953; Springfield, Ohio. pp. 32-36.

Eller, Ernest MacNeill; *The Soviet Sea Challenge;* Cowles Book Co., 1971;   8.
       Chicago.

Ellicott, John M.; *The Naval Battle of Manila;* U.S. Naval Institute Pro-   9.
       ceedings, Vol. XXVI, No. 6, Jun. 1900; Annapolis, Maryland.
       pp. 489-514.

Ellis, A. G. and Lt de vaisseau Beucker; *Bijdrage de Rennis von de*   10.
       Torpedo's of Watermijnen; L. A. Laurey, 1872; Nieuwediep. xiv,
       301 p.

Ellis, William S.; *Bangladesh: Hope Nourishes A New Nation;* National   11.
       Geographic, Vol. CXLII, No. 3, Sept. 1972; Washington, D.C.
       pp. 295-333.

"Elly Mae" One Ship - Many Jobs (USS Ellyson); All Hands, No. 481, Feb.   12.
       1957; Washington, D.C. pp. 59-63.

In England Further Trials Are Taking Place With Floating Mines; Pro-   13.
       fessional Notes, Journal of the U.S. Artillery, Vol. XXIX, No. 3,
       May-Jun. 1908; Ft. Monroe, Virginia. pg. 325.

Englehart, Alva F.; *Recent Developments in Submarine Mining;* Coast   14.
       Artillery Journal, Vol. LXXX, No. 5, Sept-Oct. 1939; Ft. Monroe,
       Virginia. pp. 232-234.

1. *The English Torpedo System;* Army and Navy Journal, Vol. XIV, No. 40, May 12, 1877; New York. pp. 646-647.

2. *En ny Opfunden Torpedo eller Mine i Vandet;* Tidsskrift for Sovaesen, Vol. XVI, 1846; Copenhagen. pg. 357.

3. Esper, George; *Blasts Rip Warship;* Washington Evening Star, Vol. CXX, No. 200, Tuesday, Jul. 19, 1972; Washington, D.C. Front page and pg. A-6.

4. *Essai de Strategie Navale;* Berger-Levrault & Cie., 1893; Paris.

5. *Études sur la Marine de Guerre. La Strategie Navale. La Tactique de Marche d'une Armée Navale;* Berger-Levrault & Cie., 1898; Paris.

6. *Evaluation of the Effectiveness of Allied Offensive Mining Operations Against Japanese Shipping in Chinese Waters;* U.S. Naval Technical Mission to Japan, Feb. 1946; S-98(N).

7. Evans, K. G.; *Design of Wooden Naval Vessels, Part I;* Shipbuilding and Shipping Record, Vol. XC, No. 12, Sept. 19, 1957; London. pp. 373-376.

8. Evans, K. G.; *Design of Wooden Naval Vessels, Part II;* Shipbuilding and Shipping Record, Vol. XC, No. 16, Oct. 17, 1957; London. pp. 507-508.

9. Evelegh, M. H.; *Mine Sweeping;* Journal of the Royal United Service Institution, Vol. LXXXVIII, No. 549, Feb. 1943; London. pp. 35-41.

10. *The Evolution of Naval Weapons;* Bureau of Naval Personnel. U.S. Government Printing Office, Mar. 1949; Washington, D.C.

11. *The Evolution of Ordnance;* Naval Ordnance Laboratory, NAVORD Report 5672, 1965; White Oak, Maryland.

12. *Examination of Torpedo Sent From the Navy Department;* Scientific American, Vol. IX, No. 15, Oct. 10, 1863; New York. pg. 229.

13. *Experiments With Torpedoes;* The Engineer, Vol. XVIII, Sept. 2, 1864; London. pg. 146.

14. *Explosion of Submarine Mines in Baltimore Harbor;* Scientific American, Vol. LXXIX, No. 14, Oct. 1, 1898; New York. pp. 218-219.

Explosion of Submarine Torpedo; Scientific American, Vol. VIII, No. 1, Jan. 3, 1863; New York. pg. 7. 1.

Fane, Francis D. and Don Moore; *The Naked Warriors;* Appleton-Century-Crofts, Inc., 1956. pp. 246-269. 2.

Fane, Robert (Pseud.); *We Clear the Way;* W. H. Allen Co., Ltd., 1943; London. 3.

Fayles, Charles Ernest; *History of the Great War. Seaborne Trade;* John Murray, 3 Volumes, 1920, 1923, 1924; London. 4.

Fayles, Charles Ernest; *Seaborne Trade;* (Official History) John Murray, 3 Volumes, 1920, 1923, 1924; London. 5.

Fein, Maier O. and Harold Masterson; *Active Sonar Display Research in the United States (U);* NUSC Technical Document 4433; Naval Underwater Systems Center, Dec. 1, 1972. CONFIDENTIAL. 6.

Ferguson, J. N.; *The Submarine Mine;* U.S. Naval Institute Proceedings, Vol. XL, Nov-Dec. 1914; Annapolis, Maryland. pp. 1697-1706. 7.

Ferraby, H. C.; *The Floating Peril;* Windsor Magazine, Vol. XLVI, No. 89, 1917; London. pp. 119-127. 8.

Ferrand, M. C.; *Torpedo and Mine Effects in the Russo-Japanese War;* U.S. Naval Institute Proceedings, Vol. XXXIII, No. 4, Dec. 1907; Annapolis, Maryland. pp. 1479-1486. 9.

Ferris, G. T.; *Our American Torpedo System at Willets Point;* Harper's Weekly, Vol. XXXII, No. 1058, Sept. 29, 1888; New York. pp. 741-744. 10.

Field, James A.; *History of United States Naval Operations, Korea;* U.S. Government Printing Office, 1962; Washington, D.C. 11.

Fighting the Submarine Menace: Barriers of Steel Across Our Harbor Entrances; Popular Science, Vol. CXL, No. 3, Mar. 1942; New York. pp. 49-51. 12.

Finding the Answer to Minesweeping; Army Navy Register, Vol. LXXVII, No. 4006, Sept. 15, 1956; Washington, D.C. pg. 6. 13.

Firing Submarine Mines by Wireless Telegraphy; Scientific American, Vol. CVII, No. 5, Aug. 3, 1912; New York. pg. 101. 14.

1. Fisher, John A.; *Handbook on Mine Warfare, Torpedo Manual, Vol. I;* Royal Navy, 1906.

2. Fisher, Richard; *With the French Minesweepers;* Selwyn and Blount, 1946; London.

3. *Fishermen Are Warned on Hauling in Derelict Mines;* Science News Letter, Vol. XLIV, Jul. 24, 1943; Washington, D.C. pg. 56.

4. *Fish-Finding From A Copter Conducted by Pye;* Journal of the American Helicopter Society, Vol. XLIV, Apr. 1957; New York. pg. 4.

5. *Fishing Up Torpedoes;* The Engineer, Vol. XXX, 1870; London. pg. 343.

6. Fiske, Bradley Allen; *Electricity in Warfare;* Scientific American Supplement, Vol. XXX, No. 768, Sept. 20, 1890; New York. pp. 1272-73.

7. Fleming, Peter; *Operation Sea Lion;* Ace Books, Inc., 1957; New York.

8. (Floating and Anchored Mines); Marine Review, Vol. XLIV, Oct. 1914; London. pg. 402.

9. *Floating Bomb Laden Nets to Foil Submarines;* Popular Mechanics, Vol. XXVIII, No. 3, Sept. 1917; Chicago. pp. 321-322.

10. *Floating Death;* Literary Digest, Vol. LX, No. 13, Mar. 29, 1919; New York. pg. 135.

11. *Floating Mine Blows Up Ship at Phnom Penh;* Washington Evening Star, Vol. CXX, No. 342, Thursday, Dec. 7, 1972; Washington, D.C. pg. A-27.

12. *Floating Mine Dangers;* Weekly Underwriter, Vol. CLVI, Jan. 4, 1947; New York. pg. 58.

13. *Floating Mines (In Sea of Japan);* Military Review, Vol. XXXVI, No. 6, Sept. 1956; Ft. Leavenworth. pg. 70.

14. *Floating Mines in the North Atlantic and Arctic Oceans;* Scientific American, Vol. CXX, No. 16, Apr. 19, 1919; New York. pp. 394-395, 416.

15. *Floating Mines in the Pacific;* Fairplay, Vol. XLVII, Sept. 6, 1906; London. pg. 359.

16. *Les Flottes de Combat en 1904 (Par le Capitaine de Fregate de Balincourt);* Berger-Levrault & Cie., Paris.

*Fluxmeter (Instrument to Measure the Strength of Magnetic Forces Remaining   1.
    in a Ship After It Has Been Commissioned);* National Defense Trans-
    portation Journal, Vol. XVI, May-Jun. 1960; Washington, D.C. pg. 16.

Fontin, Paul; *Réformes Navales;* Berger-Levrault & Cie., 1899; Paris.   2.

Footnote T; Time, Vol. XLV, No. 26, Jun. 25, 1945; New York. pg. 31.   3.

Forbes, C.; *Exploding Mines by Wireless Telegraphy;* Scientific American,   4.
    Vol. CX, No. 18, May 2, 1914; New York. pg. 371.

Forcing the Dardanelles; United Service Gazette, Vol. CLXIII, Sept. 17,   5.
    1914; London. pg. 221.

Ford, J. H.; *Harbor Obstructions and Submarine Mines;* Revue of Revues,   6.
    Vol. XVII, No. 5, May 1898; New York. pg. 593.

Foreign Notes; Newsweek, Vol. XV, No. 4, Jan. 22, 1940; New York. pg. 8.   7.

Forrest, G.; *Submarine Mine;* Scientific American, Vol. C, No. 4, Jan. 23,   8.
    1909; New York. pp. 80-81.

Forshell, Hans; *Minan I. Sjokriget;* Folkforsvaret, Farlag A.B., 1919;   9.
    Stockholm.

(Fort Totten); Army and Navy Journal, Vol. XXXIX, No. 87, May 17, 1902;   10.
    New York. pg. 924.

Fournier, F. E.; *La Flotte Nécessaire. Ses Avantages Stratégiques, Tactiques*   11.
    et Économiques; Berger-Levrault & Cie., 1896; Paris.

Le Franc, A.; *Les Mines - Ludions;* Moniteur De La Flotte, Vol. LX, No. 18,   12.
    May 3, 1913; Paris. pg. 3.

Le Franc, A.; *Les Mines Sous Marines;* Moniteur De La Flotte, Vol. LIX,   13.
    May 18, 1912; Paris. pg. 3.

Francis, D.; *Their War Is Not Over; U.S. Navy's Mine Sweepers;* Popular   14.
    Science, Vol. CXLVIII, No. 1, Jan. 1946; New York. pp. 72-76.

Fraser, I. S., <u>et al.</u>; *The Electrical Engineering Aspects of Degaussing;*   15.
    Engineer, Vol. CLXXXI, Jun. 21, 1946; London. pp. 563-565.

Fra. T.f.S. (arkivet tilstillet en Aflrandlery skrevet i 1810 om Torpedo-   16.
    Krig og Under vandesplosioner; Tidsskrift for Sovaesen, Vol. LXXXVII,
    1916; Copenhagen. pg. 588.

1. Frazier, R. W.; *The Mine Laying Career of the Royal Rajah*; National Revue, Vol. LXXII, Jun. 1919; London. pp. 563-569.

2. Freeman, Lewis R.; *Sailors Sensations in Battle; Being Torpedoed; Being Mined; State of Mind in a Naval Action;* Popular Mechanics, Vol. XXVII, No. 2, Feb. 1917; Chicago. pp. 179-183.

3. *French, English and Italian Views on Mining;* U.S. Naval Institute Proceedings, Vol. LV, No. 3, Mar. 1929; Annapolis, Maryland. pp. 237-238.

4. *French Mine, Mention of a New;* Army and Navy Journal, Vol. II, No. 46, Jul. 8, 1865; New York. pp. 723-729.

5. *French Minelayers;* U.S. Naval Institute Proceedings, Vol. XLI, No. 4, Jul-Aug. 1915; Annapolis, Maryland. pg. 1282.

6. *French Minelayers Cerbere and Pluton;* Engineer, Vol. XCVI, Nov. 14, 1913; London. pp. 515-516.

7. *The French Mine Layers Cerbere and Pluton;* Journal of the U.S. Artillery, Vol. XLI, Jan-Feb. 1914; Ft. Monroe, Virginia. pp. 361-363.

8. *French Mines;* U.S. Naval Institute Proceedings, Vol. XXXVIII, No. 2, Jun. 1912; Annapolis, Maryland. pp. 788-789.

9. *French Naval Maneuvers;* Scientific American, Vol. LV, No. 10, Sept. 4, 1866; New York. pg. 150.

10. French, W. F.; *Fishing For German Mines;* Illustrated World, Vol. XXX, No. 1, Sept. 1918; Chicago. pp. 33-38.

11. Frothingham, Thomas G.; *The Naval History of the World War, Offensive Operations 1914-1915;* Harvard University Press, 1924; Cambridge.

12. Frothingham, Thomas G.; *The Naval History of the World War, The Stress of Sea Power 1915-1916;* Harvard University Press, 1925; Cambridge.

13. Frothingham, Thomas G.; *The Naval History of the World War, The United States in the War 1917-1918;* Harvard University Press, 1926; Cambridge.

14. *Fuehrer Conferences on the German Navy 1939-1941;* Office of Naval Intelligence, U.S. Navy, 1947; Washington, D.C.

15. Fuller, J. F. C.; *Armament and History;* Charles Scribners Sons, 1945; New York.

16. Fulton, Robert; Concluding Address of Fulton's Lecture on the Mechanism Practice, Effects of Torpedoes, delivered at Washington, Feb. 17, 1810.

Fulton, Robert; *De la Machine Infernale Maritime ... Traduit de l'Anglais par M. E. Nuñez de Taboada;* Magimel, 1812; Paris. iv, 100 p.   1.

Fulton, Robert; National Cyclopedia of American Biography, Vol. III, 1893; New York. pp. 104-105.   2.

Fulton, Robert; *Torpedo War and Submarine Explosions;* William Elliot, 1810; 114 Water Street, New York.   3.

Fulton's Torpedoes; Scientific American, Vol. LXX, No. 7, Feb. 17, 1894; New York. pp. 104-105.   4.

Fuqua, S. O.; *By Products of Aerial Warfare;* Newsweek, Vol. XIV, No. 23, Dec. 4, 1939; New York. pg. 26.   5.

Fuses for Torpedoes; Engineer, Vol. XXXI, Jan. 20, 1871; London. pg. 46.   6.

The Future of Submarine Mining; Royal Engineers Journal, New Series, Vol. II, No. 2, Aug. 2, 1905; Chatham Kent. pp. 184-185.   7.

Fuses for Electronic Mines; Electronics, Vol. XIX, Oct. 1946; New York. pg. 162.   8.

Fyfe, Herbert C.; *Submarine Warfare Past & Present;* E. G. Richards, 1907; London.   9.

Gakuma, G.; *Torpedoes;* Scientific American, Vol. XXXVI, No. 22, Jul. 21, 1877; New York. pg. 337.   10.

Gale, Benjamin; *The American Turtle Built at Saybrook by David Bushnell;* Connecticut Historical Society, Collections Hartford, Vol. II, 1870. pp. 315-318, 322-323, 333-335.   11.

Galland, Adolf; *The First and the Last;* Translated by Merryn Savill, Methuen & Co., 1955 & 1970; London.   12.

Galvin, William M.; *The Northern Mine Barrage;* Sea Power, Vol. IX, No. 5, Nov. 1920; Washington, D.C. pp. 253-254.   13.

Garcia, R. C.; *Yorktown, Va. (Naval Schools, Mine Warfare) - Mine Man's Alma Mater;* All Hands, No. 481, Feb. 1957; Washington, D.C. pp. 16-19.   14.

Garner, J. W.; *Use of Submarine Mines;* American Journal of International Law, Vol. IX, No. 1, Jan. 1915; New York. pp. 86-93.   15.

1. Garvin, Richard L.; *"Antisubmarine Warfare and National Security"*; Scientific American; Vol. 227, No. 1, Jul. 1972.

2. Gaunt, Richard H.; *Navy's Mine Defense Programs Viewed with Renewed Emphasis*; Data, Vol. XI, Oct. 1966; Washington, D.C. pp. 57-58.

3. Gawn, R. W. L.; *Some Model Experiments in Connection with Mine Warfare*; Engineer, Vol. CLXXI, Apr. 26, 1946; London. pp. 386-387.

4. Gawn, R. W. L.; *Some Model Experiments in Connection with Mine Warfare*; Engineering, Vol. CLXI, May 10, 1946; London. pp. 449-450.

5. Gawn, R. W. L.; *Some Model Experiments in Connection with Mine Warfare*; Shipbuilding and Shipping Record, Vol. LXVII, No. 26, Jun. 1946; London. pp. 717-719.

6. Gawn, R. W. L.; *Some Model Experiments in Connection with Mine Warfare*; Transaction of Naval Architects, Paper No. 5, 1945; London.

7. *General Characteristics of German Navy (SVK) Influence Mines and Mine Firing Devices*; U.S. Naval Technical Mission in Europe, Technical Paper No. 467-45, Oct. 1945.

8. *German Mine Laying Cruises*; U.S. Naval Institute Proceedings, Vol. XLV, No. 6, Jun. 1919; Annapolis, Maryland. pp. 1029-1030.

9. *German Mines Blast Shipping Allies Tighten Sea Blockade*; Scholastic Magazine, Vol. XXXV, Dec. 4, 1939; Dayton, Ohio. pg. 7.

10. *German Minesweeper Herkules*; Illustrated London News, Vol. CCXLII, Jun. 8, 1963; London. pg. 895.

11. *German Naval Mines*; U.S. Naval Institute Proceedings, Vol. XLIII, No. 11, Nov. 1917; Annapolis, Maryland. pp. 2612-2613.

12. *German Submarine Activities on the Atlantic Coast of the United States and Canada*; Navy Department, Office of Naval Records and Library Historical Section, U.S. Government Printing Office, 1920; Washington, D.C.

13. *German Submarine Mine Layer U.C. 5*; Illustrated London News, Vol. CXLIX, 1916; London. pp. 127-129.

The German Submarine Mine-Layer U.C. 5; Professional Notes, Journal of   1.
    the U.S. Artillery, Vol. XLVI, No. 2, Sept-Oct. 1916; Ft. Monroe,
    Virginia. pp. 227-228.

German Submarine Mine-Layer U.C. 5; U.S. Naval Institute Proceedings,   2.
    Vol. XLII, No. 6, Nov-Dec. 1916; Annapolis, Maryland. pp. 1678-1679.

German Underwater Ordnance - Mines; U.S. Navy Department, Bureau of Ordnance,   3.
    OP 1673A, Jun. 14, 1946.

German Waters Now Clear of Mines; U.S. Naval Institute Proceedings, Vol.   4.
    XLVIII, No. 11, Nov. 1922; Annapolis, Maryland. pg. 1965.

Germany and the Submarine Mine; Navy League Journal, Vol. XI, No. 7,   5.
    Jul. 1906; London. pp. 168, 171-172.

Getler, Michael; *No Defense Seen for Mines;* Washington Post, Vol. XCV,   6.
    No. 162, Monday, May 15, 1972; Washington, D.C. pg. A-14.

Getler, Michael; *10 Ships Steam From Haiphong Despite Mines;* Washington   7.
    Post, Vol. XCVI, No. 79, Thursday, Feb. 22, 1973; Washington, D.C.
    pg. A-14.

de Ghappedelaine, G. (Translator); *Experience de Rupture en Russie;* Revue   8.
    Maritime et Coloniale, Vol. LIV, Jul. 1877; Paris. pp. 236-239.

Gibson, Charles R.; *War Inventions and How They Were Invented;* Part of   9.
    the Science for Children Series, Seeley, Service and Co., 1917;
    London.

Gibson, R. H.; *Three Years of Naval Warfare;* William Heinemann, 1918;   10.
    London.

Gibson, R. H. and Maurice Pendergast; *The German Submarine War, 1914-1918;*   11.
    R. R. Smith, Inc., 1931; New York.

Gillett, J. K. and C. A. Schoute-Vanneck; *Magnetic Mine Problem in South*   12.
    Africa; South African Institute of Electrical Engineers-Transactions,
    Vol. XXXVIII, Part 2, Feb. 1947; Johannesburg. pp. 61-82

Gillett, J. K. and C. A. Schoute-Vanneck; *Magnetic Mine Problem in South*   13.
    Africa; South African Institute of Electrical Engineers-Transactions,
    Vol. XXXVIII, Part 4, Apr. 1947; Johannesburg. pp. 137-138.

1. Gilmore, Arthur H.; *Anti-Torpedo Vessel;* Journal of the Royal United Service Institution, Vol. XIII, No. 53, 1869; London. pg. 33.

2. Ginocchetti, Angelo; *Nozoni di Storia Navale;* Licinis Coppelli, 1934; Bologna.

3. Gisborne, F. G.; *Provisional Patent Specification. Improvements in Means or Apparatus for Effecting the Explosion of Torpedoes and Other Explosive Charges. Patent 3907 of 22 December 1868;* Patent Office, by date; London.

4. Gisborne, F. G.; *Torpedoes;* The Engineer, Vol. XXVIII, Apr. 16, 1869; London. pg. 44.

5. Goncharov, L. G. and B. A. Denisov; *Use of Mines in World War I, 1914-1918;* Naval Military Publishing House, 1942; Moscow, Leningrad.

6. Goodeve, C. F.; *Battle of Scientists;* Engineer Journal, Vol. XXVIII, No. 9, Sept. 1945; London. pp. 568-569.

7. Goodeve, C. F.; *The Defeat of the Magnetic Mine;* Journal of the Royal Society of Arts, Vol. XCIV, No. 4708, Jan. 4, 1946; London. pp. 81-90.

8. Goodeve, C. F.; *Defeat of Magnetic Mine;* Shipbuilder and Marine Engine Builder, Vol. LIII, No. 442, Jan. 1946; London. pp. 16-18.

9. Goodrich, Casper Frederick; *Some Points in Coast Defense Brought Out by the War with Spain;* U.S. Naval Institute Proceedings, Vol. XXVII, No. 2, Jun. 1901; Annapolis, Maryland. pp. 223-246.

10. Goodrich, Casper Frederick; *Torpedoes - Their Disposition and Radius of Destructive Effect;* U.S. Naval Institute Proceedings, Vol. V, No. 10, 1879; Annapolis, Maryland. pp. 479-491.

11. Gordon, C. V.; *The Royal Netherlands Navy;* U.S. Naval Institute Proceedings, Vol. LXXXIV, No. 6, Jun. 1958; Annapolis, Maryland. pp. 89-101.

12. Gosse, Joseph; *Les Mines Sous-Marines ...;* F. Pichon, 1914; Paris.

13. Graf, George (Compiler); *International Law Topics and Discussions 1914;* U.S. Naval War College, U.S. Government Printing Office, 1915; Washington, D.C.

14. Graf, H.; *The Russian Navy in War & Revolution From 1914 Up to 1918;* Translated from Russian, R. Oldenbough, 1923; Munich.

Grant, Robert M.; *How America Defeated the U-Boat*; Sea Power, Vol. III, 1.
Oct. 1939; Washington, D.C. pp. 10-11, 13.

Grant, Robert M.; *Use of Mines Against Submarines*; U.S. Naval Institute 2.
Proceedings, Vol. LXIV, No. 9, Sept. 1938; Annapolis, Maryland.
pp. 1275-1279.

Grappin de revelvement des Cables Sous-Marines, Systeme Hensman; Le Génie 3.
Civil, Vol. LXV, No. 4, May 23, 1914; Paris. pg. 88.

Grasset, Albert; *La Défense des Côtes*; Berger-Levrault & Cie., 1899; Paris. 4.

Gray, Edwyn; *The Underwater War*; Scribner's, 1971; New York. 5.

Gray, Elisha; *Submarine Mines*; Electricity, Apr. 13, 1898; New York. 6.

Great Britain Invasion: Preview and Prevention; Time, Vol. XXXV, No. 23, 7.
Jun. 3, 1940; New York. pg. 27.

Green, Fitzhugh; *Our Naval Heritage*; Century Co., 1925; New York. pp. 376-388. 8.

Greenough, Ernest A.; *Planting and Raising Mines From A Scow*; Journal of 9.
the U.S. Artillery, Vol. XLVI, Jul-Dec. 1916; Ft. Monroe, Virginia.
pp. 60-63.

Grenfell, Russell; *The Art of the Admiral*; Faber & Faber, Ltd., 1937. 10.
London.

Gretton, Peter W.; *The British Navy in the Spanish Civil War*; Manuscript, 11.
personal communication.

Griswold, Charles; *Submarine Navigation*; American Journal of Science and 12.
Arts, Vol. II, 1820; New Haven, Connecticut. pp. 94-100.

Grivel, R.; *La Marine et les Bombardements des Villes du Littoral:* 13.
Sebostopol, Bomarsund-Odessa-Sweaborg Kinburn; 1856; Paris.

Grosjean, M.; *La Mine Sous-Marine*; L'Europe Nouvelle, Vol. XXIII, Apr. 20, 14.
1940; Paris. pg. 436.

Grosjean, M.; *Mouillage et Draguage de Mines*; L'Europe Nouvelle, Vol. 15.
XXIII, Apr. 13, 1940; Paris. pg. 399.

Grosvenor, Joan and L. M. Bates; *Open the Ports (The Story of Human Mine-* 16.
sweepers); William Kimber, 1956; London.

Groves, Donald; *Helos Revolutionize Minesweeping*; Our Navy Magazine, 17.
Vol. LXVII, No. 3, Mar. 1972; Brooklyn, New York.

1. *La Guerre avec l'Angleterre. Politique Navale de la France;* Berger-Levrault
    & Cie., 1899; Paris.

2. *La Guerre de Mines;* L'Illustration, Vol. CCIV, No. 5043, Oct. 28, 1939;
    Paris. pp. 214-215.

3. Guichard, Louis; *The Naval Blockade 1914-1918;* Translated by Christopher
    R. Turner, Phillip Allan and Co., Ltd., 1930; London.

4. Guierre, A.; *L'Avenir de la Torpille et la Guerre Future;* Berger-Levrault
    & Cie., 1898; Paris.

5. Gwynne, Alban Lewis; *The Modern Submarine Mine;* Journal of the Royal United
    Service Institution, Vol. LXXIV, Nov. 1929; London. pp. 807-811.

6. Gwynne, Alban Lewis; *The Submarine Mine;* Brassey's Annual for 1929,
    Chapter VIII, Wm. Clowes & Son, 1919; London. pp. 106-123.

7. Gwynne, Alban Lewis; *The Submarine Mine as a Naval Mine;* Vickers Co., Ltd.,
    1933; London.

8. Hahn, Fritz; *Deutches Geheimwaffen: 1939-1945;* Erich Hoffman Verlag, 1963;
    Heidenheim.

9. Hailey, Foster; *Mines Still Floating in Area Off Japan;* U.S. Naval Institute
    Proceedings, Vol. LXXXII, No. 7, Jul. 1956; Annapolis, Maryland.
    pp. 779-780.

10. Hailey, Foster and Lancelot Milton; *Clear for Action (The Photographic
    Story of Modern Naval Combat 1898-1964);* Duell, Sloan and Pearce,
    1964; New York.

11. Haldane, J. B. S.; *Magnetic Mines;* New Statesman and Nation, Vol. XVIII,
    No. 458, Dec. 2, 1919; London. pp. 781-782.

12. Hall, Charles H.; *Mines and Torpedoes;* Sea Power, Vol. II, No. 4, Apr.
    1917; Washington, D.C. pp. 40-44.

13. Hall, Cyril; *Modern Weapons of War by Land, Sea and Air;* Blackie & Son,
    Ltd., 1915; London.

14. Hall, J. A.; *The Law of Naval Warfare;* Chapman & Hall, Ltd., 1921; London.

15. Halsey, F. W.; *The Literary Digest History of the World War;* Funk and
    Wagnall's Company, 1920; New York and London.

Hamilton, John Randolph; *The American Navy*; Journal of the Royal United   1.
    Service Institution, Vol. XII, No. 59, 1869; London. pp. 243-271.

Hampshire, A. Cecil; *International Minesweeping Comes to an End*; The   2.
    Navy, Vol. LVII, No. 3, Mar. 1952; London. pp. 62-63.

Hampshire, A. Cecil; *Minesweeping Tops the Bill*; The Navy, Vol. LVII,   3.
    No. 10, Oct. 1952; London. pp. 284-285.

Handbook for Seaman Gunners; U.S. Naval Torpedoes Station, Newport,   4.
    Rhode Island. U.S. Government Printing Office, 1908; Washington,
    D.C.

von Handel-Mazzetti, P. A.; *Österreich-Ungarns Flotte im Weltkrieg*;   5.
    1924; Innsbruck-Vienna.

Handle with Care (Training at the Mine Warfare School at Yorktown, Va.);   6.
    All Hands, No. 490, Nov. 1957; Washington, D.C. pg. 15.

Hanks, Carlos C.; *Mines of Long Ago*; Coast Artillery Journal, Vol. LXXXIII,   7.
    No. 6, Nov-Dec. 1940; Ft. Monroe, Virginia. pp. 539-541.

Hanks, Carlos C.; *Mines of Long Ago*; U.S. Naval Institute Proceedings,   8.
    Vol. LXVI, No. 11, Nov. 1940; Annapolis, Maryland. pp. 1548-1551.

Hansards; *Parliamentary Debates, June 5, 1810*; Vol. XVII, 2nd Series,   9.
    pp. 305-306.

Hansen, H.; *Sominevaesenets Historie*; Tidsskrift for Sovaesen, Vol. XCIX,   10.
    Apr. 1928. pg. 155.

Harbor Defense; Army and Navy Journal, Vol. XI, No. 25, Jan. 31, 1874;   11.
    New York. pg. 392.

(Harbor Defence); *The Army and Navy Gazette Says*; Army and Navy Journal,   12.
    Vol. XXVIII, No. 6, Oct. 4, 1890; New York. pg. 93.

(Harbor Defense); *Foreign Items*; Army and Naval Journal, Vol. XIV, No. 36,   13.
    Apr. 14, 1877; New York. pg. 579.

Harbour Defence. One View of Abolition of Mines; The Navy, Vol. X, No. 11,   14.
    Nov. 1905; London. pp. 286-287.

Hare, Robert; *Application of Galvanic Ignition to Rock Blasting*; American   15.
    Journal Science and Arts, B. Sulliman, Editor, Vol. XXI, Jan. 1832;
    New Haven, Connecticut. pp. 139-141.

1.   Hare, Robert; *On a Calorimotor for Igniting Gases in Eudiometrical Experiments and Gunpowder in Rock-Blasting*; Notices and Abstracts of Communications to the British Association for the Advancement of Science at the Bristol Meeting, Aug. 1836, Vol. V, 1837; London. pg. 45.

2.   Harriman, J. E.; *Submarine Mines; Their Purpose, Characteristics, and Operation*; Ordnance, Vol. XX, No. 119, Mar-Apr. 1939; Washington, D.C. pp. 306-308.

3.   Harris, Robert Hastings; *The Changes in the Conditions of Naval Warfare*; Journal of the Royal United Service Institution, Vol. XXX, No. 134, 1886; London. pp. 437-440.

4.   Hartley, A. B.; *Unexploded Bomb*; W. W. Norton & Co., Inc., 1958; New York.

5.   Hartmann, Gregory K.; *U.S. Naval Ordnance Laboratory - 50 Years of Research*; U.S. Naval Institute Proceedings, Vol. XCV, No. 10, Oct. 1969; Annapolis, Maryland. pp. 138-141.

6.   Harvey, John; *On the Construction of Vessels in Relation to the Changed Modes of Naval Warfare*; Engineer, Vol. XXXIII, Mar. 29, 1872; London. pg. 221.

7.   Harvey, John; *On the Construction of Vessels in Relation to the Changed Modes of Naval Warfare*; Transactions of the Institute of Naval Architects, Vol. XIII, 1872; London. pp. 3-5.

8.   Hauck, Russel; *British Mine Countermeasures (1914-15) and the Dardanelles Operation*; U.S. Naval Schools, Mine Warfare, Term Paper No. 55, Nov. 21, 1971; Charleston, South Carolina.

9.   Haven, Charles T. and Frank A. Belden; *A History of the Colt Revolver*; William Morrow & Co., 1940; New York.

10. Hayes, John D. (Editor); *The Civil War Correspondence of S. F. DuPont, U.S.N.*; Cornell University Press, 3 volumes, 1969; Ithaca, New York.

11. "H.B."; *Apparat til overskaering of Mineslryningstrosser*; Tidsskrift for Sovaesen, Vol. LXXXVI, 1915; Copenhagen. pg. 411.

12. "H.B."; *Apparat til Udlaegning of undersoiske Miner eller Stromminer*; Tidssdrift for Sovaesen, Vol. LXXXVI, 1915; Copenhagen. pg. 467.

13. Heald, Joseph F.; *Mine Naval*; Encyclopedia International, Vol. XII, Grolier, Inc., 1964. pp. 103-104.

Heinl, Robert D., Jr.; *The Inchon Landing: A Case Study in Amphibious Planning;* Naval War College Review, Vol. XIX, No. 9, May 1967; Newport, Rhode Island. pp. 51-72.   1.

Heinl, Robert D., Jr.; *Reasons for Mining NVN Waters;* Armed Forces Journal, Vol. CIX, No. 10, Jun. 1972; Washington, D.C. pg. 62.   2.

Helicopter Clears Minefield; Army and Navy Register, Vol. LXXVI, No. 3929, Mar. 26, 1955; Washington, D.C. pg. 23.   3.

Helicopter Minesweeper; Military Review, Vol. XXXV, No. 5, Aug. 1955; Ft. Leavenworth. pg. 67.   4.

(A helicopter tows a magnetic hydrofoil minesweeping device); Photo, National Observer, Vol. XII, No. 7, Feb. 17, 1973; Silver Spring, Maryland. pg. 2.   5.

Hennebert, Eugene; *Les Torpilles;* Librairie Hachette et Cie., 1884; Paris.   6.

Hennebert, Eugene; *Les Torpilles par le Major H. de Sarrepont;* J. Dumaine, 1st Edition, 2nd Edition, 1874, 1888; Paris.   7.

Henningsen, Charles Fredrick; *Memorial, Respectfully Addressed to the Congress of Confederate States;* 1864; Richmond.   8.

Herrick, Robert Waring; *Soviet Naval Strategy; Fifty Years of Theory and Practice;* United States Naval Institute, 1968.   9.

Hershey, Amos S.; *The International Law and Diplomacy of the Russo-Japanese War;* Macmillan, 1906; New York. pp. 124-125.   10.

Hessler, William H.; *Sweden's Armed Neutrality;* U.S. Naval Institute Proceedings, Vol. LXXXI, No. 1, Jan. 1955; Annapolis, Maryland. pp. 39-49.   11.

Higgins, Alexander Pearce; *The Hague Peace Conferences and Other International Conferences Concerning the Laws and Usages of War Texts of Conventions With Commentaries;* Cambridge University Press, 1909.   12.

Highest Court; Time, Vol. LIII, No. 16, Apr. 18, 1949; New York. pg. 31.   13.

Hine, Al; *D-Day the Invasion of Europe;* American Heritage Publishing Co., Inc., 1962; New York.   14.

1. Hinkamp, Clarence Nelson; *Bring in the Sheaves;* U.S. Naval Institute Pro-
        ceedings, Vol. XLV, No. 7, Jul. 1919; Annapolis, Maryland. pp. 1117-
        1133.

2. Hinkamp, Clarence Nelsor *Mine Sweeping;* Sea Power, Vol. VII, No. 4, Oct.
        1919; Washington, .C. pp. 188-191.

3. Hinkamp, Clarence Nelson; *Pipe Sweepers;* U.S. Naval Institute Proceedings,
        Vol. XLVI, No. 9, Sept. 1920; Annapolis, Maryland. pp. 1477-1484.

4. *His Majesty's Minesweepers;* Prepared for the Admiralty by the Ministry of
        Information, 1943; London.

5. *Histoire des Torpilles;* J. Dumaine, 1877; Paris. 42 p. [Brochure]

6. *History of Switch Horn Development at NOL;* Naval Ordnance Laboratory,
        H 148, Apr. 18, 1946; White Oak, Maryland.

7. Hoblitzell, James J.; *The Lessons of Mine Warfare;* U.S. Naval Institute
        Proceedings, Vol. LXXXVIII, No. 12, Dec. 1962; Annapolis, Maryland.
        pp. 32-37.

8. Hobson, Richmond P.; *The Sinking of the "Merrimac";* The Century Magazine,
        Vol. LVII, No. 3, Jan. 1899; New York. pp. 427-450.

9. Hoehling, Adolph A.; *The Great War at Sea - A History of Naval Action
        1914-18;* Crowell Co., 1965; New York.

10. Hoehling, Adolph A.; *Wooden Warships Are Back Again;* American Mercury,
        Vol. LVII, Oct. 1953; New York. pp. 97-99.

11. Holman, H. R.; *Harbor Defense Torpedoes;* Scientific American, Vol. VI,
        No. 3, Jan. 18, 1862; New York. pg. 38.

12. Holmes, Nathaniel J.; *The Application of Electricity as a Means of Defence
        in Naval and Military Warfare;* Journal of the Society of Telegraph
        Engineers, Vol. III, 1874; London. pp. 32-51.

13. Holmes, Nathaniel J.; *Military Torpedo Defences;* Journal of the Society
        of Telegraph Engineers, Vol. III, 1874; London. pp. 54-79.

14. Holmes, Nathaniel J.; *Torpedo Defences;* Journal of the Royal United Service
        Institution, Vol. X, No. 38, May 1866; London. pp. 402-414.

15. *Holmes Torpedo Finder;* Engineering, Vol. XXXIII, Apr. 7, 1882; London.
        pp. 343, 357.

Kolmes, Wilfred J.; *Undersea Victory, The Influence of Submarine Operations on the War in the Pacific*; Doubleday, 1966; Garden City, New York.　1.

Holt, David H.; *The Anguish of Normandy*; Mil Review, Vol. XLV, No. 6, Jun. 1965; Ft. Leavenworth. pp. 56-62.　2.

Honours for Mine-Sweepers; The Navy, Vol. XX, No. 3, Mar. 1915; London. pp. 82-83.　3.

Hornsnaill, W. O.; *War Beneath the Waves*; Chambers Journal, Series 7, Vol. V, 1915; London. pp. 293-294.　4.

Houllevigue, L.; *Mines et Torpilles Sous-Marines*; Revue de Paris, Vol. XXII, No. 2, 1915; Paris. pp. 344-366.　5.

How Britain Was Fed in Wartime: Food Control 1939-1945; Great Britain Ministry of Food, Her Majesty's Stationery Office, 1946; London.　6.

How Mines are Laid and Fired; Popular Mechanics Magazine, Vol. XXVI, No. 1, Jul. 1916; Chicago. pp. 68-69.　7.

How Mines Help Guard America's Harbors; Popular Mechanics, Vol. LXXIV, No. 6, Dec. 1940; Chicago. pg. 813.　8.

How "Smart Bombs" Are Squeezing North Vietnam; U.S. News and World Report, Vol. LXXII, No. 23, Jun. 5, 1972; Washington, D.C. pp. 23-24.　9.

Hubbard, J. C.; *Future Use of Submarines*; U.S. Naval Institute Proceedings, Vol. LXII, No. 12, Dec. 1936; Annapolis, Maryland. pp. 1721-1726.　10.

Hubbard, Miles H.; *Naval Ordnance Heritage*; Ordnance, Vol. XLIV, No. 236, Sept-Oct. 1959; Washington, D.C. pp. 222-224.　11.

Hubbe, M.; *The Luy Moveable Torpedo*; W. W. Rowley, 1880; Buffalo, New York.　12.

Huber, James; *Submarine Mines*; Engineering Magazine, Vol. L, Oct. 1915; New York. pp. 120-121.　13.

Huet, C.; (Défense des Rades - 1875) Les Mines Sous-Marines dans la Défense des Rades. Note sur un Mécanisme Automoteur Noyé Obligeant les Torpilles a suivre le Mouvement des Deniselations de la Marée; Bruxelles, Muqardt et Paris, Dumaine, 1875. 48 p.　14.

1. Hughes, Hobart; *Saga of a YMS*; U.S. Naval Institute Proceedings, Vol. LXXIV, No. 1, Jan. 1948; Annapolis, Maryland. pp. 53-59.

2. Hull, E. W. Seabrook; *Asleep in the Deep?*; Ordnance, Vol. LXIX, No. 266, Sept-Oct. 1964; Washington, D.C. pp. 158-161.

3. Hurd, A. S.; *The Merchant Navy*; (Official History), Her Majesty's Stationery Office, 5 volumes, 1921-1929; Vol. I, 1921; London.

4. Hurd, Archibald; *Italian Sea Power in the Great War*; Constable and Co. Ltd., 1918; London.

5. Hurwitt, Albert; *The Navy's Role in the Occupation of Japan*; Sea Power, Vol. VI, No. 11, Nov. 1946; Washington, D.C. pp. 35-37, 39.

6. Husnu, Kurt; *Mine Warfare Operations During World War I at Dardanelles in Turkey*; U.S. Naval Schools, Mine Warfare, Senior Foreign Officer Course 7001, Paper No. 6, Jun. 1970.

7. Hyman, Jerome; *The Scrub Women*; (Mine Sweepers); Sea Power, Vol. IV, No. 10, Oct. 1944; Washington, D.C. pp. 10-11, 29.

8. *Illegal Mining and Bomb-Dropping*; Scientific American, Vol. CXI, No. 12, Sept. 19, 1914; New York. pg. 222.

9. *Illustrative Studies in Mine Warfare*; U.S. Naval War College, 1955; Newport, Rhode Island.

10. *Improvements in and Connected With Submarine Mines*; Engineer, Vol. CXXI, No. 3313, Jan. 14, 1916; London. pg. 50.

11. *Incidents in Corfu Channel*; Security Council Considers British Charges Against Albania, United Nations Bulletin, Vol. II, Mar. 4, 1947; New York. pp. 178-184.

12. *Infernal Machines in the Mississippi*; Scientific American, Vol. VI, No. 14, Apr. 5, 1862; New York. pp. 210-211.

13. *Information Annual: A Digest of Current Events, 1915*; R. R. Bowker Co., 1916; New York.

14. *Instructions for the Management of Harvey's Sea Torpedo*; E. and F. N. Spon, 1871; London.

15. *International Law Studies*; U.S. Naval War College, 1935; Newport, Rhode Island. pp. 298, 682.

Interior Protection Torpedoes and Mines; Journal of the U.S. Artillery,   1.
    Vol. XLII, No. 3, May-Jun. 1914; Ft. Monroe, Virginia. pp. 358-361.

In the Air: To Keep Afloat; Time, Vol. XXXV, No. 4, Jan. 22, 1940;   2.
    New York. pg. 29.

Ipsen, P.; *On Anuendelsen af forankreden Miner;* Besvarelse of   3.
    Solojtnantsselskabets Presspørgsmall, Tidsskrift for Sovaesen,
    Vol. XCIII, 1922.

Iron Men at Work; Navy, Vol. V, No. 8, Aug. 1962; Washington, D.C.   4.
    pg. 13.

Iron Sea Monsters; All Hands, No. 493; Feb. 1958; Washington, D.C.   5.
    pg. 22.

Irwin, A. E.; *Story of the Mine;* The Sea Cadet, Vol. I, No. 8, 1944;   6.

Italian Mines in the Mediterranean; Time, Vol. XXXV, No. 26, Jun. 24, 1940;   7.
    New York. pg. 27.

The Italian Navy in the World War, 1915-1918; Proveditorato Generale   8.
    Dello Stato Libreria, Office of the Chief of Staff of the Royal
    Italian Navy, 1927; Rome.

Italian Navy's Mine Layers; U.S. Naval Institute Proceedings, Vol. LIII,   9.
    No. 2, Feb. 1927; Annapolis, Maryland. pg. 357.

Italian Submarine Mines; Scientific Australian, Vol. XXIII, No. 1,   10.
    Sept. 1917; Melbourne. pp. 13-14.

The Italian-Turkish War; U.S. Naval Institute Proceedings, Vol. XXXVIII,   11.
    No. 3, Sept. 1912; Annapolis, Maryland. pg. 1069.

"J.A.K."; *Krigsskibes Beskyttelse mod Mine-og Torpedospraengninger;*   12.
    Tidsskrift for Sovaesen, Vol. LXXXVII, 1916; Copenhagen. pg. 622.

"J.A.K."; *Sikkerhedsanordning til forankrede undersoiske miner;*   13.
    Tidsskrift for Sovaesen, Vol. LXXXVII, 1916; Copenhagen. pg. 150.

(Jacobi); Soviet Encyclopedia, Vol. LI. pp. 521-523.   14.

Jane, Fredrich T.; *Defense of Harbors Against Torpedo Boat Attack;*   15.
    Journal of the U.S. Artillery, Vol. XXII, No. 3, Nov-Dec. 1904;
    Ft. Monroe, Virginia. pp. 187-191.

1. Jane, Fredrich T.; *The Imperial Russian Navy*; W. Thacker & Co., 1899; London.

2. *Jane's Weapons Systems 1972-73*; McGraw-Hill, 1972; New York.

3. *Jap War Echoes*; Newsweek, Vol. XXX, No. 24, Dec. 15, 1947; Washington, D.C. pg. 24.

4. *Japanese to Have Minesweeping Helicopter Squadron by 1972*; U.S. Naval Institute Proceedings, Vol. XCVI, No. 7, Jul. 1970; Annapolis, Maryland. pp. 132-133.

5. Jaques, William Henry; *Torpedoes for National Defence*; G. P. Putnam's Sons, 1886; New York.

6. Jeffers, William N.; *Harbor Defense*; Army and Navy Journal, Vol. XI, No. 26, Feb. 7, 1874; New York. pg. 410.

7. Jellicoe, John Rushworth; *The Crisis of the Naval War*; Cassell and Co., Ltd., 1920; London, New York, Toronto and Melbourne.

8. Jellicoe, John Rushworth; *The Grand Fleet 1914-15*; George H. Doran Co., 1919; New York.

9. Jervois, W. F. Drummond; *Coast Defences and the Application of Iron to Fortification*; Journal of the Royal United Service Institution, Vol. XII, No. 52, 1868; London. pp. 549-569.

10. *John Lewis Lay*; National Cyclopedia American Biography, Vol. VII, 1897. pp. 528-529.

11. Johns, A. W.; *German Submarines*; Engineering, Vol. CIX, No. 2830, Mar. 26, 1920; London. pp. 428-432.

12. Johnson, Bruce H.; *Mine Warfare Operations in the Russo-Japanese War (1904-1905)*; U.S. Naval Schools, Mine Warfare, Staff Officers Course Term Paper, Nov. 25, 1968; Charleston, South Carolina.

13. Johnson, Ellis A. and David A. Katcher; *Mines Against Japan*; Report to the Navy, Naval Ordnance Laboratory, 1947; White Oak, Maryland. CONFIDENTIAL. [Unclassified version, U.S. Government Printing Office, 1973.]

14. Johnson, John; *The Defense of Charleston Harbor*; Walker, Evans and Cogswell Co., 1890; Charleston, South Carolina.

15. Johnston, Oswald; *U.S. Resumes Mining of Haiphong Harbor: Diplomatic Pressure*; Evening Star, Vol. CXX, No. 353, Monday, Dec. 18, 1972; Washington, D.C. pp. A-1, A-6.

Joint Army-Navy Assessment Committee; *Japanese Naval and Merchant Shipping*   1.
Losses During World War II by All Causes; U.S. Government Printing
Office, Feb. 1947; Washington, D.C.

The Jones Buoyant Torpedo Guard; Marine Engineer, Vol. XXII, Oct. 1, 1901;   2.
London. pp. 282-284.

Jones, C. B.; *The Use of Mines by Enemy Seen as Real Threat in Viet Nam;*   3.
Navy, Vol. IX, No. 6, Jun. 1966; Washington, D.C. pp. 14-15.

Jones, Henry L.; *History of Army Mine Planters;* Coast Artillery Journal,   4.
Vol. LXXX, No. 5, Sept-Oct. 1939; Ft. Monroe, Virginia. pp. 456-458.

Jones, H. Robert; *Mine Countermeasures;* U.S. Naval Schools, Mine Warfare,   5.
Jan. 4, 1968; Charleston, South Carolina.

Jones, J. M.; *The Vanguard Had to be Minecraft;* U.S. Naval Schools, Mine   6.
Warfare, Staff Officers Course Term Paper, Nov. 3, 1969; Charleston,
South Carolina.

Jones, J. William; *Southern Historical Society Papers;* Vol. II, George   7.
W. Gray, printer and stationer, 1876; Richmond, Virginia.

Jones, Virgil Carrington; *U.S.S. Cairo;* National Park Service, U.S. Dept.   8.
of Interior, U.S. Government Printing Office, 1971; Washington, D.C.
pp. 28, 30-31.

Jones, Virgil Carrington; *The Civil War at Sea;* Holt, Rinehart, Winston,   9.
Vol. III, 1962; New York.

Jones, Virgil Carrington; *The Civil War at Sea: The Final Effort;* Holt,   10.
Rinehart, Winston, 1962; New York.

Jones, Virgil Carrington; *The Civil War at Sea: The River War;* Holt,   11.
Rinehart, Winston, 1962; New York.

Jose, A. W.; *Official History of Australia in the War of 1914-1918;* Vol.   12.
IX, The Royal Australian Navy; Angus and Robertson, 1928; Sydney.

Julicher, Peter J.; *Mine Warfare Operations in the Dardanelles Campaign*   13.
of 1914-1915; U.S. Naval Schools, Mine Warfare, Staff Officers Course
Term Paper, Jun. 1969; Charleston, South Carolina.

Juul, C.; *Søminelaere;* 1st Edition, 1878, 1880, 1882 (three sections).   14.

1. Kalmpffert, W.; *Those Magnetic Mines*; Science Digest, Vol. VII, No. 2, Feb. 1940; Chicago. pp. 33-35.

2. Kannengiesser, Hans Pasha; *The Campaign in Gallipoli*; Hutchinson and Co., Ltd., 1928; London.

3. Karig, Walter Cagle and F. A. Manson; *Battle Report of the War in Korea*; Rinehart and Co., 1952.

4. Karl, R. L. and J. H. Thorton, Jr.; *Nonmagnetic Minesweepers*; Ordnance, Vol. XXXIX, No. 208, Jan-Feb. 1955; Washington, D.C. pp. 657-660.

5. Karneke, Joseph Sidney and Victor Boesen; *Navy Diver*; Putnam, 1962; New York.

6. Kauffman, Draper L.; *German Naval Strategy in World War II*; U.S. Naval War College, for the Strategy and Tactics Dept., 1951; Newport, Rhode Island.

7. Kay, Howard N.; *Reenlistment: A Case History*; U.S. Naval Institute Proceedings, Vol. LXXXII, No. 6, Jun. 1956; Annapolis, Maryland. pp. 602-609.

8. Kay, Howard N.; *Scientific Wonderland*; U.S. Naval Institute Proceedings, Vol. LXXVIII, No. 4, Apr. 1952; Annapolis, Maryland. pp. 389-397.

9. Keate, E. M. and W. C. B. Tunstall; *A Short Naval Bibliography*; C. Knight and Co., 1926; London.

10. Kekewich, Piers K.; *The Use of Submarine Mines as Affected by the VIIIth Hague Convention*; Translated from an article by G. Larghezza in Revista Marittima, Journal of the Royal United Service Institution, Vol. LIX, No. 438, Nov. 1914; London. pp. 455-468.

11. Kelly, H. W. K.; *Degaussing*; Nature, Vol. CLVIII, No. 3994, May 18, 1946; London. pp. 646-648.

12. Kelly, H. W. K.; *Research in Countering Magnetic Mine, History*; Electrician, Vol. CXXXVI, No. 3541, Apr. 12, 1946; London. pp. 953-954.

13. Kelly, Orr; *POW's Due Next Week: Mine Removal Set*; Evening Star, Vol. CXXI, No. 26, Friday, Jan. 26, 1973; Washington, D.C. pg. A-1.

14. Kiep, U. H. A.; *"Die Seemine" Die Technik im Zwanzisten Jahrhundert*; Vol. VI, 1916; Braunschweig. pp. 284-288.

15. Kim, Sang Mo; *The Implications of the Sea War in Korea*; Naval War College Review, Vol. XX, Summer 1967; Newport, Rhode Island. pp. 105-139.

King, Cecil; *Atlantic Charter*; Studio Publication, 1943; London and New York. 1.

King, Cecil; *Rule Britannia*; Studio Publication, 1941; London and New York. 2.

King, Ernest J.; *U.S. Navy at War 1941-1945*; U.S. Navy Department, 1946; 3.
    Washington, D.C.

King, Ernest J. and Walter M. Whitehill; *Fleet Admiral King: A Naval Record*; 4.
    Norton and Co., 1952; New York.

King, William Rice; *Submarine Warfare*; Army and Navy Journal, Vol. VII, 5.
    No. 13, Nov. 13, 1869; New York. pg. 195.

King, William Rice; *Torpedoes: Their Invention and Use*; U.S. Army Corps of 6.
    Engineers, 1866; Washington, D.C.

Kingsford, H.; *Electrical Grapnel for Submarine Cables and Torpedo Lines*; 7.
    Scientific American Supplement, Vol. XVII, No. 430, Mar. 29, 1884;
    New York. pg. 6859.

Kinney, Sheldon; *All Quiet at Wonsan*; U.S. Naval Institute Proceedings, 8.
    Vol. LXXX, No. 8, Aug. 1954; Annapolis, Maryland. pp. 859-867.

Kirby, S. Woodburn; *The War Against Japan, Vol. III: The Decisive Battles*; 9.
    Her Majesty's Stationery Office, 1961; London.

Kirk, John and Robert Young, Jr.; *Great Weapons of World War II*; Walker 10.
    and Co., 1961; New York.

Klado, Nicholas L.; *La Marine Russe dans la Guerre Russo-Japonaise*; Berger- 11.
    Levrault & Cie., 1905; Paris.

Klado, Nicholas L.; *The Battle of the Sea of Japan*; Translated by J. H. 12.
    Dickinson and F. P. Marchant, Hodder and Stoughton, 1906.

Knapp, H. S.; *Results of Some Special Researches at the Torpedo Station*; 13.
    U.S. Naval Institute Proceedings, Vol. XIX, No. 3, 1893; Annapolis,
    Maryland. pp. 249-266.

Knox, Dudley W.; *A History of the United States Navy*; G. P. Putnam's Sons, 14.
    1936; New York.

Knox, Thomas W.; *Robert Fulton*; G. P. Putnam's Sons, 1900; New York. 15.

Kolbenschlog, George R.; *Minesweeping on the Long Tao River*; U.S. Naval 16.
    Institute Proceedings, Vol. XCIII, No. 6, Jun. 1967. pp. 88-102.

1. Korotkin, I. M.; *Battle Damage to Ships During World War II*; Sunpromgiz, 1960; Leningrad.

2. Krafft, Herman F. and Walter B. Norris; *Sea Power in American History*; Century, 1920; New York.

3. Kraus, J. H.; *Why An Acoustic Mine Explodes: How To Make A Working Model*; Science News Letter, Vol. XL, Nov. 8, 1941; Washington, D.C. pg. 296.

4. *Kriegswerth der Seeminen und Torpedoes, ueber den*; Fr. Wilh. Grunow, 1881; Leipzig. 64 p.

5. Kvam, Kare E.; *Minekrig Til Sjøs*; Olof Hanssens Fond. Horten, Norway.

6. Lake, Simon; *Submarines That Are Strictly Invisible*; Scientific American, Vol. CXII, No. 3, Jan. 16, 1915; New York. pp. 68-69, 74-75.

7. Lamont, R. R.; *War Under the Waves Modern Mines and Minesweeping*; The Sea Cadet, Vol. III, No. 12, 1946; London.

8. Langmaid, Kenneth J. R.; *The Approaches Are Mined!*; Jarrolds, 1965; London.

9. Lankford, B. W. and J. E. Pinto; *Developments in Wooden Minesweeper Hull Design Since World War II*; Naval Engineers Journal, Vol. LXXIX, Apr. 1967; Washington, D.C. pp. 275-289.

10. Lankford, B. W. and J. F. Anglrer; *Glass Reinforced Plastics Developments for Application to Minesweeper Construction*; Naval Engineers Journal, Vol. LXXXIII, Oct. 1971; Washington, D.C. pp. 13-26.

11. *Largest Mine Field in History*; Army and Navy Journal, Vol. LVI, No. 9, Nov. 2, 1918; New York. pg. 315.

12. Larghezza, G.; *L'Uso delle Mine nella Guerra Marittima*; Revista Marittima, Vol. XLVII, No. 1, Feb. 1914; Rome. pp. 234-252.

13. Laubeuf, Alfred Maxime and Henri Stroh; *Sous-Marines, Torpilles et Mines*; Librarie J. B. Bailierre et Fils, 1923; Paris.

14. (Launching of the French Mine Laying Cruiser Emile Bertin, May 9, 1933); Marine Engineering and Shipping Age, Vol. XXXVIII, No. 8, Aug. 1933; New York. pg. 285.

15. Lauth, J.; *L'État Militaire des Principales Puissances Étrangeres en 1902*; Berger-Levrault & Cie., Paris.

-55-

The Law of War and Neutrality at Sea; U.S. Government Printing Office, 1957;　1.
　　Washington, D.C. pg. 303.

Lawrence, T. J.; *War and Neutrality in the Far East;* MacMillan and Co.,　2.
　　Ltd., 1904; London. pp. 93-111.

Lebedskoi, G. M.; *Torpedoes and Mines;* Military P lishing House, 1949;　3.
　　Moscow.

Ledieu, Alfred; *Le Noveau Material Naval par A. Ledieu et Ernest Cadiat;*　4.
　　Vve C. Donod, 1889-90; Paris.

Ledig, Gerhard; *Bilderbuch der Minensuch;* Hase & Koehler, 1943; Leipzig.　5.

Legrand, J.; *La Leçon de Fashoda;* Berger-Levrault & Cie., 1899; Paris.　6.

Lepotier, Capt. de Vais A.; *Mer Contre Terre - Les Leçons de l'Histoire*　7.
　　(1861-1865); Mirambeau, 1945; Paris. 358 p.

The Leon Torpedo; The Engineer, Vol. CXX, Aug. 27, 1915; London. pp. 196-　8.
　　197.

The Leon Torpedo; Journal of the Royal Society of Arts, Vol. LXIII, 1915;　9.
　　London. pp. 1031-1032.

The Leon Torpedo; Marine Engineer, Vol. XXXVIII, Part V, No. 459, Dec. 1915;　10.
　　London. pg. 115.

Leon Torpedo; New York Times, Oct. 2, 1915. pg. 3.　11.

Leon Torpedo and Detect...; New York Times, Part I, Oct. 3, 1915. pg. 3　12.

Leon Torpedo-Mine; U.S. Naval Institute Proceedings, Vol. XLI, No. 3,　13.
　　May-Jun. 1915; Annapolis, Maryland. pp. 942-944.

Letter From Paris; Mines, The Navy, Vol. V, No. 7, Jul. 1911. pp. 9-11.　14.

Levie, Howard S.; *Mine Warfare and International Law;* Naval War College　15.
　　Review, Vol. XXIV, No. 8, Apr. 1972; Newport, Rhode Island.
　　pp. 27-35.

Lewis, Charles Lee; *David Glasgow Farragut, Our First Admiral;* U.S. Naval　16.
　　Institute, 2 volumes, 1941, 1943; Annapolis, Maryland.

1. Lewis, Charles Lee; *Mathew Fontaine Maury, Pathfinder of the Sea;* U.S. Naval Institute, 1927; Annapolis, Maryland.

2. Lewis, Charles Lee; *The Romantic Decatur;* University of Pennsylvania Press, 1937; Philadelphia, Pennsylvania.

3. Lewis, Michael; *The Navy in Transition 1814-1864;* A Social History; Hodder and Stoughton, 1965; London.

4. Lewis, Michael; *The Navy of Britain;* George Allen and Unwin, 1947; London.

5. *Lidt om KMA-Miner og Forstrondsminer Samt Rydningen of disse;* Tiddskrift for Sovaesen, Vol. CXVII, 1946; Copenhagen. pg. 271.

6. *Life's Photographer Sweeps on British Trawler in North Sea;* Life, Vol. VIII, Jan. 8, 1940; Chicago. pp. 18-19.

7. *Lightning-Effect on Controlled Mines;* Army and Navy Journal, Vol. XII, No. 11, Oct. 24, 1874; New York. pg. 175.

8. Liman von Sanders, O. V. K.: *Five Years in Turkey;* U.S. Naval Institute, 1927; Annapolis, Maryland.

9. Lincoln, Fredman Ashe; *Secret Naval Investigator;* William Kimber, 1961; London.

10. Lissak, Ormond M.; *Ordnance & Gunnery;* John Wiley & Sons, Inc., 1915; New York.

11. *Liste du Matériel de Torpilles;* Autographie du Ministère de la Marine, 1875; Paris. 27 p. [Atlas]

12. *Liste de Tous les Objets Composant le Matériel Actuel de Torpilles;* Imprimerie Nationale, 1875; Paris. 127 p.

13. *Live Mines at Canal No Danger to Navigation;* Science News Letter, Vol. XXXVIII, Jul. 6, 1940; Washington, D.C. pg. 8.

14. Lloyd, Christopher (Editor); *The Keith Papers, Vol. III 1803-1815;* Printed for the Navy Record Society, 1955.

15. *Location and Repair of Faults in Submarine Mine Cable;* U.S. Coast Artillery School, Coast Artillery School Press, 1913; Ft. Monroe, Virginia.

Lochner, R.; *Backroom in Battle Dress; The Fight Against Magnetic Mines;* Blackwoods Magazine, Vol. CCLXI, No. 1578, Apr. 1947; London. pp. 348-360.   1.

Lockroy, Edouard; *La Défense Navale;* Berger-Levrault & Cie., 1899; Paris.   2.

Lockroy, Edouard; *La Marine de Guerre;* Berger-Levrault & Cie., 1897; Paris.   3.

Lockroy, Edouard; *Du Weser à la Vistule. Lettres sur la Marine Allemande;* Berger-Levrault & Cie., 1901; Paris.   4.

Lohman, Walter; *Kamaraden auf See Zwischen Minen und Torpedoes;* 1943; Berlin.   5.

Lohr, Carl A.; *The Principles Involved in the Mine Defense of Harbors;* Journal of the U.S. Artillery, Vol. XLV, No. 3, May-Jun. 1916; Ft. Monroe, Virginia. pp. 229-313.   6.

London Naval Treaty of 1936; U.S. Department of State, Conference Series No. 24.   7.

Long, A.; *Navy Foils Magnetic Mine;* Science News Letter, Vol. LXIV, Aug. 22, 1953; Washington, D.C. pp. 123-124.   8.

Looking for Dynamite; Newsweek, Vol. XXVII, No. 2, Jun. 11, 1946; Washington, D.C. pp. 56-57.   9.

Looking for Trouble; All Hands, Navy Bureau of Personnel, No. 358, Dec. 1946; Washington, D.C. pp. 28-29.   10.

Loosbrock, J. F.; *Mines Are Dirty Tricks;* Popular Science, Vol. CLVII, No. 2, Feb. 1951; New York. pp. 107-112.   11.

Lorey, Hermann; *Der Krieg in den Turkischen Gewassern;* Mittler und Sohn, Vol. I, 1928; Vol. II, 1938; Berlin.   12.

Lossing, Benson J.; *The Pictorial Field Book of the War of 1812;* Harper Brothers, 1869; New York.   13.

Lossing, Benson J.; *Pictorial History of the Civil War in the United States of America;* Volume I by George W. Childs, 1866; Philadelphia. Volumes II and III by T. Belknap, 1866; Hartford.   14.

Lott, Arnold S.; *Japan's Nightmare-Mine Blockade;* U.S. Naval Institute Proceedings, Vol. LXXXV, No. 11, Nov. 1959; Annapolis, Maryland. pp. 39-51.   15.

1. Lott, Arnold S.; *Most Dangerous Sea*; G. Banta, 1959; Menosha, Wisconsin. xiv, 322 p.

2. Lott, Arnold S.; *Most Dangerous Sea*; U.S. Naval Institute, 1959; Annapolis, Maryland.

3. Lott, Davis Newton; *Sweeping to Victory*; Sea Power, Vol. IV, No. 4, Apr. 1944; Washington, D.C. pp. 22-23, 25.

4. Lott, Davis Newton; *United States Wooden Ships and Green Men Are Making A Name For Themselves*; Yachting, Vol. LXXVI, Jul. 1944; New York. pp. 36-38, 105.

5. Loughton, L. G.; *War Under Water*; Monthly Review, Vol. XVI, London. pg. 60.

6. Loukine, Alexandre; *The Submarine Minelayer Krab*; U.S. Naval Institute Proceedings, Vol. LX, Feb. 1934; Annapolis, Maryland. pp. 197-201.

7. Low, Archibald Montgomery; *Mine and Countermine*; Sheridan House, 1940; New York.

8. Low, Archibald Montgomery; *Warfare's Deadly Mines*; Science Digest, Vol. IX, No. 4, Apr. 1941; Chicago. pp. 9-15.

9. Lukin, A. P.; *Secrets of Mine Warfare*; U.S. Naval Institute Proceedings, Vol. LXVI, No. 5, May 1940; Annapolis, Maryland. pp. 642-643.

10. Lundeberg, Philip K.; *Undersea Warfare and Allied Strategy in World War I, Part I: To 1916*; The Smithsonian Journal of History, Vol. I, No. 3, 1966; Washington, D.C.

11. Lundeberg, Philip K.; *Undersea Warfare and Allied Strategy in World War I; Part II: 1916-1918*; The Smithsonian Journal of History, Vol. I, No. 4, Winter, 1967; Washington, D.C.

12. Lupinacci, Pier Fillippo and Vittorio Tognelli; *La Guerra de Mine*; Vol. XVIII, La Marina Italiana Nella Seconda Guerra Mondiale, 1966; Rome.

13. Lusar, Rudolph; *Weapons and Secret Weapons of World War II and Their Subsequent Development*; J. I. Lehmanns Verlag, 1956; Munich.

14. "M"; *Einiges über Schiffsartillerie, Torpedoes und Seeminen in der Franzosischen Marine*; Marine Rundschau, Vol. XVI, 1905; Berlin, pp. 407-415.

"M.A.W".; *Mine-Sweeping in the North Sea*; The Field, Vol. CXXV, 1915; London.   1.
      pg. 90.

McClintock, Robert; *The River War in Indochina*; U.S. Naval Institute Pro-   2.
      ceedings, Vol. LXXX, No. 12, Dec. 1954; Annapolis, Maryland. pp. 1303-
      1318.

McClung, Frank; *Roto-Craft to the Sweep*; United Aircraft Beehive, Vol.   3.
      XLII, Fall 1967; East Hartford, Connecticut. pp. 12-15.

MacDonald, H.; *Mine Sweepers*; Chambers Journal, Vol. V, No. 234, May 22,   4.
      1915; Edinburgh. pp. 385-389.

MacDonald, H.; *Minesweepers*; Living Age, Vol. CCLXXXVI, Aug. 21, 1915;   5.
      Boston. pp. 473-476.

McEarthron, Ellsworth D.; *Minecraft in the Van*; U.S. Government Printing   6.
      Office, 1956; Washington, D.C.

McElgin, Hugh J. B.; *Suggestions for the Organization of Work in a Large*   7.
      Mine Command; Journal of the U.S. Artillery, Vol. XLII, No. 3,
      Nov-Dec. 1914; Ft. Monroe, Virginia. pp. 301-311.

McEntee, Gerard Lindsley; *Italy's Part in Winning the World War*; Princeton   8.
      Press, 1934.

McEntee, Gerard Lindsley; *Military History of the World War*; Charles   9.
      Scribners Sons, 1931; New York.

McGrath, Thomas D.; *Mines Vs. Submarines*; Ordnance, Vol. LII, No. 280,   10.
      Jan-Feb. 1967; Washington, D.C. pp. 393-395.

McGrath, Thomas D.; *The Submarine's Long Shadow*; U.S. Naval Institute Pro-   11.
      ceedings, Vol. XCII, No. 6, Jun. 1966; Annapolis, Maryland. pg. 117.

MacGregor, Edgar John; *Sea Communications World War III*; Research Paper   12.
      for the Air University, 1952; Maxwell Field, Alabama.

Machiavelli, Niccolo; *The Art of War*; Review Edition of the Ellis Farnworth   13.
      Translation; Bobbs Merrill, 1965.

Machiavelli, Niccolo; *A Gentleman of the State New York*; Hints relative   14.
      to torpedo warfare and addition to *The Art of War in Seven Books*.
      Published by Henry C. Southwich, 1815; Albany, New York.

1. McIlwraith, Charles G.; *The Mine as a Tool of Limited War*; U.S. Naval
    Institute Proceedings, Vol. XCIII, No. 6, Jun. 1967; Annapolis,
    Maryland. pg. 103.

2. MacIntyre, Donald; *A Forgotten Campaign*; Journal of the Royal United
    Service Institution, Vol. CVI, Feb., May, Aug. and Nov. 1961;
    London. pp. 65-70, 242-247, 405-409 and 554-560.

3. MacIntyre, Donald; *The Naval War Against Hitler*; Scribner, 1971; New York.

4. McKay, Robert F.; *The Paravane*; Engineering, Vol. CVIII, No. 2803,
    Sept. 19, 1919; London. pp. 389-392.

5. McKay, Robert F.; *Paravanes*; Engineer, Vol. CXXXII, No. 3418, Jul. 1, 1921;
    London. pp. 3-4.

6. McLachlon, Bruce; *They Also Serve*; Navy, Vol. V, No. 8, Aug. 1962; Washington,
    D.C. pp. 16-17.

7. Maclean, Alstair S.; *They Sweep the Seas*; Chambers Journal, 8th Series,
    Vol. X, No. 596, Sept. 1941; Edinburg. pp. 561-564.

8. McMurtrie, Francis; *Minelaying in the War*; The Sea Cadet, Vol. III,
    1946.

9. de Maconge, J. L; *Une Marine Rationelle. La Flotte Utile. Les Réformes
    Necessaires de Notre Organisme Naval*; Berger-Levrault & Cie.,
    1903; Paris.

10. *Magnetic Mine: History of Degaussing*; Electrician, Vol. CXXXV, No. 3512,
    Sept. 21, 1945; London. pp. 302-303.

11. *Magnetic Mine Sweeper*; Ordnance, Vol. LVII, No. 313, Jul-Aug. 1972;
    Washington, D.C. pg. 76.

12. *Magnetic Mines and Buoyant Cable*; The Engineer, Vol. CLXXIX, No. 4665,
    Jun. 8, 1945; London. pp. 447-448.

13. *Magnetic Mines Feasible Say American Experts*; Science News Letter,
    Vol. XXXVI, Dec. 2, 1939; Washington, D.C. pg. 358.

14. *Magnetic Mines Still Being Found*; Marine Engineering, Vol. LIII, No. 10,
    Oct. 1948; New York. pp. 62-63.

15. *Magnetic Torpedoes*; Scientific American, Vol. VI, No. 15, Apr. 19, 1862;
    New York. pg. 230.

Magneto-Electric Firer; The Engineer, Vol. XXXIV, Aug. 2, 1872; London.   1.
    pg. 69.

Magruder, T. P.; *The Navy in the War;* The Saturday Evening Post, Vol. CCI,   2.
    No. 39, Mar. 30, 1929; Philadelphia, Pennsylvania. pp. 20-21, 117,
    121.

Magruder, T. P.; *The Navy in the War;* The Saturday Evening Post, Vol. CCI,   3.
    No. 41, Apr. 13, 1929; Philadelphia, Pennsylvania. pp. 29, 129-130,
    133, 137.

Mahan, Alfred Thayer; *La Guerre sur Mer et Ses Leçons;* (Guerre Hispano-   4.
    Americaine - 1898); Berger-Levrault & Cie., 1899; Paris.

Mahan, Alfred Thayer; *Naval Strategy Compared and Contrasted with the*   5.
    Principles and Practice of Operations on Land; Little and Brown,
    1911; Boston.

Mahan, Alfred Thayer; *The Navy in the Civil War: Admiral Farragut;*   6.
    Appleton, 1897; New York.

Mahan, Alfred Thayer; *The Navy in the Civil War: The Gulf and Inland*   7.
    Waters; Sampson Low, Marston, 1898; London.

Mahan, Alfred Thayer; *Sea Power in its Relation to the War of 1812;*   8.
    Sampson Low, Marston, 1905; London.

Malcolm, E. D.; *Military Submarine Mining;* Blackwood, Vol. CLXXVIII,   9.
    No. 573, Aug. 1905; London. pp. 288-292.

Mallot, Robert; *Subaqueous Torpedoes;* Naval Science, Vol. I, Jul. 1872;   10.
    London.

Manfroni, C.; *Storia della Marine Italiana durante la Guerra Mondiale,*   11.
    1914-1918; 1923; Bologna.

Mann, C. F. A.; *Gas Turbines Are Here;* Diesel Progress, Vol. XX,   12.
    No. 9, Sept. 1954; New York. pp. 25-27.

Mann, C. F. A.: *Minesweeper AM 435 Powered with New Packard Diesel;*   13.
    Diesel Progress, Vol. XX, No. 7, Jul. 1954; New York. pp. 32-43.

Mann, C. F. A.; *New Navy Mine Sweeper;* Diesel Progress, Vol. XX,   14.
    No. 4, Apr. 1954; New York. pp. 37-39.

1. Mann, C. F. A.; *Wood and Diesel Go to War as YMS 241 Joins "400"*; Diesel Progress, Vol. IX, No. 4, Apr. 1943; New York. pp. 42-45.

2. *Manners & Morals (Americana)*; Time, Vol. L, No. 21, Nov. 24, 1947; New York. pg. 30.

3. Mannix, D. P.; *Raiders of the Night*; Saint Nicholas Magazine, Vol. LVII, Aug. 1930; Columbus, Ohio. pp. 762-765.

4. *Manual of Submarine Mining*; Compiled by order of HRH the Field Marshal Commanding in Chief, HMSO, Harrison and Sons, 1880; London.

5. *Manuel des Défenses Sous-Marines*; Dumaine et Imprimerie Nationale, 1873 (-1886); Paris. [Électricité et torpilles.]

6. Marder, Arthur J.; *The Anatomy of British Sea Power. A History of British Naval Policy in the Pre-dreadnought Era 1880-1905*; Knopf, 1940; New York.

7. Marder, Arthur J.; *From Dreadnought to Scapa Flow*; Ebenezer Baylis and Son, Ltd., 1961; London.

8. Marder, Murrey; *Haiphong: U.S. Should Clear Mines*; Washington Post, Vol. XCVI, No. 48, Monday, Jan. 22, 1973; Washington, D.C. Front page and pg. A-12.

9. *La Marina Italiana nella Guerra Europea 1916-1918*; Ministero della Marina, Nov. 1916-1918; Milano. [Eleven chapters published at various times.]

10. *La Marina Italiana nella Seconda Guerra Mondiale*; Officio storico della marina militare, 1950 Vol. XVII, La Guerra de Mine, Compilatore: Cap. de Vasc. (r.) Pier Filippo Lupinacci, 1966; Roma. 487 p.

11. *Marine Mines (Their Purpose: How They Are Planted and How Destroyed)*; Scientific American Supplement, Vol. LXXX, No. 2085, Dec. 18, 1915; New York. pp. 388-389.

12. *Mark Versus Mine*; OP 572; U.S. Government Printing Office, 1915; Washington, D.C.

13. Marks, E.; *Fighting Magnetic Mines*; Sheet Metal Industries, Vol. XIV, No. 155, Mar. 1940; London. pp. 265-268.

Marshall, H. W. S.; *Some Notes on Mines and Mine Sweeping;* South African   1.
    Institute of Electrical Engineers-Transactions, Vol. XXXVIII,
    Part 2, Feb. 1947; Johannesburg. pp. 50-61.

Martienssen, Anthony; *Hitler and His Admirals;* E. P. Dutton & Co., 1949;   2.
    New York.

Martin, L. W.; *The Seas in Modern Strategy;* The Institute for Strategic   3.
    Studies, 1968; Fredrich Praeger.

Masterman, J. C.; *The Double-Cross System in the War of 1939 to 1945;*   4.
    Yale University Press; 1972.

Material and Manual for Submarine Mining for the Engineer Service; The   5.
    Chief of Engineers, U.S. Army, Mar. 24, 1887; Willets Point, New
    York Harbor.

Material of the Submarine Mining Service of the United States with a Manual   6.
    for its Use in Coast Defense; 3rd Edition, Battalion of Engineers
    Press, 1898; Willets Point, New York.

(Maury, Matthew Fontaine); National Cyclopedia of American Biography;   7.
    Vol. VI, 1896; New York. pp. 35-36.

Maury, Richard Lancelot; *A Brief Sketch of the Work of Matthew Fontaine*   8.
    Maury During the War 1861-65; Whitlet and Shepperson, 1915;
    Richmond, Virginia.

Mayo, Claude Banks; *Your Navy;* Parker & Baird, 1943; Los Angeles,   9.
    California.

Meacham, James Alfred; *Four Mining Campaigns: An Historical Analysis of*   10.
    the Decisions of the Commanders; Naval War College Review, Vol.
    XIX, No. 10, Jun. 1967; Newport, Rhode Island. pp. 75-129.

Meacham, James Alfred; *The Mine as a Tool of Limited War;* U.S. Naval   11.
    Institute Proceedings, Vol. XCIII, No. 2, Feb. 1967; Annapolis,
    Maryland. pp. 50-62.

Meacham, James Alfred; *The Mine Countermeasures Ship;* U.S. Naval Institute   12.
    Proceedings, Vol. XCIV, No. 4, Apr. 1968; Annapolis, Maryland.
    pp. 128-129.

Meacham, James Alfred; *Whatever Became of the Mine;* U.S. Naval Institute   13.
    Proceedings, Vol. XCII, No. 3, Mar. 1966; Annapolis, Maryland. pp.
    115-117.

Mechanical Torredoes; Engineering, Vol. XXIII, May 25, 1877; London. pp.   14.
    406-408.

1. *Mechanical Torpedoes;* Engineering, Vol. XXIII, Jun. 25, 1877; London. pg. 456.

2. *Medd. fra Sominevaesenets Omraade, Med den Russisk-Japanske Krig som Baggrund;* Tidsskrift for Sovaesen, Vol. LXXVII, 1906; Copenhagen.

3. Medlicott, W. N.; *The Economic Blockade;* History 2nd World War U.K. Civil Series, 2 Volumes, HMSO, 1952; London.

4. Meister, Jurg; *Russian Mine Warfare;* Navy, the Magazine of the British Navy League, Vol. LXII, No. 6-12 and Vol. LXIII, No. 1-6. (Russian Mine Warfare, Vol. LXII, No. 9, Sept. 1957. pp. 293-295.)

5. Meister, Jurg; *Der Seekrieg in den Osteurpaischen Gewassern 1941-1945;* Lehmann, 1958; München.

6. de Mello Tamborim, A. J.; *The Paraguayan Torpedoes;* Army and Navy Journal, Vol. XI, No. 41, May 23, 1874; New York. pp. 650-651.

7. de Mello Tamborim, A. J.; *The Torpedo War in Paraguay;* Army and Navy Journal, Vol. XII, No. 10, Oct. 17, 1874; New York. pg. 154.

8. *Memoirs: Brigadier General Sir James Edmonds;* Royal Engineers Journal, Vol. LXX, No. 4, Dec. 1956; Chatham Kent. pp. 395-398.

9. *Memorandum on the Effectiveness of Mines Laid by Aircraft in the Blockade of England;* German Naval Staff, 1941.

10. *The Menace of the Mine;* Navy and Army Illustrated, Vol. I, New Series, 1914; London. pp. 14-16.

11. *Menace of the Seas;* Time, Vol. XLIX, No. 4, Jan. 27, 1947; New York. pg. 26.

12. Mercer, David D.; *The Baltic Sea Campaign 1918-20;* U.S. Naval Institute Proceedings, Vol. LXXXVIII, No. 9, Sept. 1962; Annapolis, Maryland. pp. 64-69.

13. Merrifield, C. W.; *The Effects of Torpedoes on Naval Construction;* Engineer, Vol. XXXIII, Mar. 29, 1872; London. pg. 226.

14. Merrifield, C. W.; *The Effects of Torpedoes on Naval Construction;* Engineering, Vol. XIII, Mar. 1872; London. pp. 198-199, 212, 213.

15. Merrifield, C. W.; *The Effect of Torpedoes on Naval Construction;* Journal of the Royal Society of Arts, Vol. XX, 1872; London. pp. 363-365.

Merrifield, C. W.; *The Effects of Torpedoes on Naval Construction;* 1.
    Transactions of the Institute of Naval Architects, Vol. XIII,
    1872; London. pp. 6-12.

Merrifield, C. W.; *The Effects of Torpedoes on Naval Construction;* 2.
    Van Nostrand's Engineering Magazine, Vol. VII, No. 43, Jul. 1872,
    New York. pp. 137-140.

Message to Nixon: U.S. Moves are Blunting Hanoi's Drive; U.S. News 3.
    and World Report, Vol. LXXII, No. 22, May 29, 1972; Washington,
    D.C. pp. 33-34.

Method for Removing Submerged Torpedo; Scientific American, Vol. X, 4.
    No. 4, Jun. 23, 1864; New York. pg. 60.

Method of Applying Torpedoes for Harbor Defense; Scientific American, 5.
    Vol. XI, No. 1, Jul. 2, 1864; New York. pg. 10.

Methods of Mine Sweeping; Chambers Journal, 7th Series, Vol. V, 6.
    No. 257, Oct. 30, 1915; Edinburgh. pp. 761-762.

Michaels, R.; *Mine Picker for Cargo Ships;* Illustrated World, Vol. XXIX, 7.
    No. 3, May 1918; Chicago. pg. 360.

Michel, N. B.; *Ship Board Degaussing Installations for Protection Against* 8.
    Magnetic Mines; Electrical Engineering, Vol. LXVIII, No. 10292,
    Jan. 1949; New York. pg. 15.

Mielichhofer, Sigmund; *Sea Coast Artillery and Submarine Mine Defense;* 9.
    Journal of the U.S. Artillery, Vol. VI, No. 3, Nov-Dec. 1896;
    Ft. Monroe, Virginia. pp. 348-359.

The Mighty Midgets; All Hands, No. 511, Aug. 1959; Washington, D.C. 10.
    pp. 10-11.

Milbury, C. E.; *Mystery of the Magnetic Mine;* Scientific American, 11.
    Vol. CLXXI, No. 3, Mar. 1940; New York. pp. 256-257.

Miles, A. H.; *Navy Mine Depot, Yorktown, Virginia;* U.S. Naval Institute 12.
    Proceedings, Vol. 54, No. 4, Apr. 1928; Annapolis, Maryland. pp.
    299-304.

Military Secret: *Self-Buoyant Electrical Cable Which Neutralized the Magnetic* 13.
    Mine; Canadian Mining Journal, Vol. XLVI, Dec. 1945; Gardenvale,
    Quebec. pp. 912-913.

1. Miller, Henry A.; *Measurement of Magnetic Fields Beneath Ships;* Journal of the Royal Society of Arts, Vol. XCIV, No. 4711, Apr. 1946; London. pp. 327-329.

2. Miller, Michael; *The Minesweeping/Fishing Vessel;* U.S. Naval Institute Proceedings, Vol. CXVI, No. 11, Nov. 1970; Annapolis, Maryland. pp. 82-83.

3. Miller, Richards T.; *Minesweepers;* Naval Review, U.S. Naval Institute, 1967; Annapolis, Maryland.

4. Miller, W. J.; *Little Ships Clear Way for Big Ships;* All Hands, No. 429, U.S. Naval Bureau of Personnel, 1952; Washington, D.C. pp. 2-7.

5. Millholland, Ray; *The Splinter Fleet of the Otranto Barrage;* The Bobs-Merrill Company, 1936; New York.

6. Milligan, John D.; *Gunboats Down the Mississippi;* U.S. Naval Institute, 1965; Annapolis, Maryland.

7. *Mine and Countermines U.S.N.;* Bureau of Ordnance, Prepared at the Naval Torpedo Station, 1902; Newport, Rhode Island.

8. *Mine Attacks Lighthouse;* Time, Vol. XXXVIII, No. 24, Dec. 15, 1941; New York. pp. 31-32.

9. *Mine Barrage;* Published by the New York Chapter North Sea Mine Force Association, undated.

10. *Mine Barrier From Norway to Scotland;* Revue of Revues, Vol. LXIX, No. 5, May 1919; New York. pp. 537-539.

11. *The Mine Carrying Ships "Cerbere" and "Pluton";* U.S. Naval Institute Proceedings, Vol. XL, No. 1, Jan-Feb. 1914; Annapolis, Maryland. pp. 197-198.

12. *Mine Casualties Continue at Frequent Rate;* National Underwriter, Vol. LIII, No. 5, Feb. 3, 1949; Chicago. pg. 4.

13. *Mine Clearing Talks Begin in Haiphong;* Evening Star, Vol. CXXI, No. 36, Monday, Feb. 5, 1973; Washington, D.C. Front page.

14. *Mine Control Protects Neutral Shipping;* Popular Mechanics, Vol. XXVI, 1916; Chicago. pp. 481-482.

15. *Mine Defense of Our Coasts;* Journal of the U.S. Artillery, Vol. XXXIV, No. 1, Jul-Aug. 1910; Ft. Monroe, Virginia. pg. 110.

Mine Defences of Our Coasts; United Service Gazette, Jul. 1, 1909; London. 1.

(Mine Design); The Engineer; Vol. CLXXX, Nov. 1945; London. 2.

Mine Explodes by Sound From Ship's Engines; Popular Mechanics, Vol. LXXVII, 3.
   No. 2, Feb. 1942; Chicago. pg. 33.

Mine-Field; Blackwoods Magazine, Vol. CCV, No. 739, Jan. 1919; London. 4.
   pp. 34-45.

Minefields Lengthen the Allied Supply Route to Finland; Newsweek, Vol. XV, 5.
   No. 3, Jan. 15, 1940; New York. pp. 19-20.

The Mine Force Family; All Hands, No. 481, Feb. 1957; Washington, D.C. 6.
   pg. 24.

Mine Killers at Work; Popular Mechanics, Vol. LXXX, No. 5, Nov. 1943; 7.
   Chicago. pp. 78-79.

Minekrigen; Tidsskrift for Sovaesen, Vol. LXXXVIII, 1917; Copenhagen. 8.
   pg. 123.

Minelayers; U.S. Naval Institute Proceedings, Vol. LII, No. 8, Aug. 1926; 9.
   Annapolis, Maryland. pg. 1585.

Mine Layers Cerbere and Pluton; International Marine Engineering, Vol. 10.
   XVIII, No. 9, Sept. 1913; New York. pg. 375.

Minelayers: Historical Transactions 1893-1943; Society of Naval Architects 11.
   and Marine Engineers, 1945; New York. pg. 324.

Minelaying; U.S. Naval Institute Proceedings, Vol. LIV, No. 6, Jun. 1928; 12.
   Annapolis, Maryland. pp. 494-495.

Mine Laying and Torpedo Regulation Vessel for Portugal; Engineering, Vol. 13.
   XC, Sept. 9, 1910; London. pp. 362-363.

Mine Laying and Torpedo Regulating Vessel; Marine Review, Vol. XL, Oct. 14.
   1910; London. pg. 389.

A Mine-Laying Squadron; Journal of the U.S. Artillery, Vol. XXXVIII, 15.
   Jul-Aug. 1912; Ft. Monroe, Virginia. pp. 115-116.

A Mine-Laying Squadron; U.S. Naval Institute Proceedings, Vol. XXXVIII, 16.
   No. 3, Sept. 1912; Annapolis, Maryland. pg. 1128.

1. *Mine-Laying Steamer for Portugal;* Journal of the U.S. Artillery, Vol. XXXIII, No. 2, Mar-Apr. 1910; Ft. Monroe, Virginia. pp. 217-218.

2. *Mine-Laying Submarine Latest in War Craft;* Popular Mechanics, Vol. LXX, No. 5, Nov. 1938; Chicago. pp. 698-699.

3. *Mine-Laying Vessel;* U.S. Naval Institute Proceedings, Vol. XXXVIII, No. 4, Dec. 1912; Annapolis, Maryland. pg. 1660.

4. *Minelaying Vessels;* U.S. Naval Institute Proceedings, Vol. XLVII, No. 11, Nov. 1921; Annapolis, Maryland. pp. 1803-1804.

5. *Mine Materials;* Army and Navy Journal, Vol. XI, No. 22, Jan. 10, 1874; New York. pg. 342.

6. *Mine, Minesweeping and Mineclearing;* Encyclopedia Americana, Vol. XIX, 1968; New York. pp. 153-154.

7. *Minengefährdete Gebiete der Europaischen Gewasser und Zwangwege der Ost- und Nordsee;* Seehydrographischer Dienst, Democratic Republic of Germany, 1962; Rostock.

8. *The Mine Planter;* Army and Navy Journal, Vol. LXVI, No. 43, Jun. 26, 1909; New York. pg. 1215.

9. *Mine Planter Ellery W. Niles;* Marine Engineering and Shipping Review, Vol. LXIII, No. 3, Mar. 1938; New York. pp. 108-115.

10. *Mine Planter Ellery W. Niles;* Marine Engineering and Shipping Review, Vol. XLIII, No. 11, Nov. 1938; New York. pg. 515.

11. *Mine Planting by Aircraft (From the N.Y. Herald Tribune, Nov. 25, 1939);* U.S. Naval Institute Proceedings, Vol. LXVI, No. 1, Jan. 1940; Annapolis, Maryland. pp. 136-138.

12. *Mines;* National Observer, Vol. XI, No. 21, May 20, 1972; Silver Spring, Maryland.

13. *Mines;* Newsweek, Vol. XVII, No. 9, Mar. 3, 1941; New York. pp. 19-21.

14. *Mines;* U.S. Naval Institute Proceedings, Vol. XXXVIII, No. 3, Sept. 1912; Annapolis, Maryland. pg. 1111.

15. *Mines and Minesweeping;* Chambers' Encyclopedia, Vol. IX, New Edition, Newnes Limited, 1959; London. pp. 427-429.

Mines and Mine-Sweeping; The Navy, Vol. XX, No. 8, Aug. 1915; London.   1.
    pg. 257.

Mines and Seamen; The Times Literary Supplement, No. 2010, Aug. 10, 1940;   2.
    London. pg. 383.

Mines and Subterranean Torpedoes at Port Arthur; Scientific American Sup-   3.
    plement, Vol. LIX, No. 1516, Jan. 21, 1905; New York. pg. 24289.

Mines (at Sea) Information Annual 1915; R. R. Bowker and Co., 1916;   4.
    New York. pp. 389-390.

Mines Cleared From the North Sea; U.S. Naval Institute Proceedings, Vol.   5.
    XCVII, No. 1, Jan. 1971; Annapolis, Maryland. pg. 108.

Mines Cleared in North Sea; Christian Science Monitor, Oct. 29, 1970;   6.
    Boston.

Mines et Poseurs de Mines de Blocus; Journal de Marine, le Yacht, Vol.   7.
    XXXV, No. 1783, May 11, 1912; Paris. pp. 289-290.

Mines Floating Towards Neutral Coasts; Literary Digest, Vol. LIII, No. 2,   8.
    Jul. 8, 1916; New York. pg. 63.

Mines - How They Work; Navy, Vol. V, No. 8, Aug. 1962; Washington, D.C.   9.
    pp. 14-15.

Mines in Naval Warfare; Naval and Military Record, Aug. 19, 1914; London.   10.

Mines in Naval Warfare (Their Use in the Russo-Japanese Struggle);   11.
    Scientific American Supplement, Vol. LXXVIII, No. 2021, Sept. 26,
    1914; New York. pg. 198.

Mines in the North Sea; Scientific American, Vol. CXI, Dec. 12, 1914;   12.
    New York. pg. 489.

Mines in the Open Sea; The Navy, Vol. XI, No. 7, Jul. 1906; London. pp.   13.
    171-172.

Mines Laid During Wars Still Shipping Menace; Weekly Underwriter, Vol.   14.
    CLXXII, No. 13, Mar. 26, 1955; New York. pg. 789.

(Mines on crusiers and battleships); United Service Gazette, Oct. 19, 1911;   15.
    London.

1. *Mines Reported Self-Exploding in Viet Waters*; Evening Star, Vol. CXXI, No. 76, Saturday, Mar. 17, 1973; Washington, D.C. pg. A-11.

2. *Les Mines Sous-Marines Russes et Japonaises*; Journal de Marine, le Yacht, Vol. XXVII, No. 1425, Jul. 1, 1905; Paris. pp. 404-405.

3. *Mines Still Floating in Area Off Japan*; New York Times, Mar. 18, 1956; New York.

4. *Mine Subacquee, Loro Importanza, Modo di Impiegarle e di Combatterle*; Rivista de Artiglieria E. Genio, Vol. IV, Oct. 1898; Rome. pp. 143-147.

5. *Mine Submarine*; New Funk and Wagnalls Encyclopedia, Vol. XXIII, Unicorn Publishing Co., 1952; New York. pp. 8492-8493.

6. *Mine, Submarine*; New International Encyclopedia, Dodd, Mead and Co., Vol. XV, 1916; New York. pp. 713-715.

7. *Minesweeper*; Engineering, Vol. CLXXXVI, No. 4842, Dec. 26, 1958; London. pg. 833.

8. *Mine Sweeper Attached to Ships*; U.S. Naval Institute Proceedings, Vol. XLIII, No. 8, Aug. 1917; Annapolis, Maryland. pg. 1803.

9. *Mine Sweeper Diesels*; Baldwin, Vol. I, No. 4, 4th Quarter, 1944; Philadelphia. pp. 20-22.

10. *Mine Sweeper Launched in Japan*; Military Review, Vol. XXXVI, No. 4, Jul. 1956; Ft. Leavenworth. pg. 74.

11. *Minesweeper with Cyclodial Propellers*; Motorship, Vol. XXXI, No. 7, Jul. 1946.

12. *Minesweeper with Cyclodial Propellers*; U.S. Naval Institute Proceedings, Vol. LXXII, No. 8, Aug. 1946; Annapolis, Maryland. pp. 1121-1123.

13. *Mine Sweepers Go in First*; Armed Forces Chemical Journal, Vol. V, No. 2, Oct. 1951; Washington, D.C. pp. 18-23, 46.

14. *Minesweepers: LVTs, to be Built for Navy*; Army and Navy Journal, Vol. LXXXVIII, No. 16, Dec. 16, 1950; Washington, D.C. pg. 415.

15. *Minesweepers with Roll Stabilization*; Engineering, Vol. CLXXXVI, No. 4842, Dec. 26, 1958; London. pg. 833.

-71-

Mine Sweeping; Living Age, Vol. CCC, Mar. 1, 1919; Boston. pp. 532-540.   1.

Mine Sweeping; U.S. Naval Institute Proceedings, Vol. XLII, No. 4, Jul-Aug. 1916; Annapolis, Maryland. pg. 1311.   2.

Mine Sweeping by Helicopter; Air Pictorial, Vol. XVII, May 1955; London. pg. 153.   3.

Mine Sweeping Devices; Army and Navy Journal, Vol. LIII, No. 34, Apr. 22, 1916; New York.   4.

Mine Sweeping Experiences; U.S. Naval Institute Proceedings, Vol. XLIII, No. 6, Jun. 1917; Annapolis, Maryland. pg. 1343.   5.

Minesweeping in Haiphong Harbor; Crash, Fire Mar; Evening Star, Vol. CXXI, No. 79, Tuesday, Mar. 20, 1973; Washington, D.C. pg. A-7.   6.

Mine Warfare - Past and Present; Navy, Vol. V, No. 8, Aug. 1962; Washington, D.C. pp. 6-12.   7.

Mine Warfare School Moves to Charleston; Air Force Times, Vol. XIX, Jan. 17, 1959; Washington, D.C. pg. E-4.   8.

Mine Work at Sea; U.S. Naval Institute Proceedings, Vol. XXXII, No. 2, Jun. 1906; Annapolis, Maryland. pp. 722-723.   9.

Mines Swept From Channel Near Albania; U.S. Naval Institute Proceedings, Vol. LXXIII, No. 1, Jan. 1947; Annapolis, Maryland. pg. 106.   10.

Mines Throttled Japanese Shipping; Sea Power, Vol. V, No. 11, Nov. 1945; Washington, D.C. pg. 56.   11.

Mining by Submarines; Army and Navy Gazette, Jul. 29, 1916; London. pg. 483.   12.

Mining Operations of German Submarines Around the British Isles, 1915-1918; Intelligence Division Naval Staff, O.U. 6333, NID 4039/39.   13.

Mining the High Seas; The Navy, Vol. IX, No. 6, Jun. 1904; London. pg. 176.   14.

Mining Their Own Business; All Hands, No. 549, Oct. 1962; Washington, D.C. pp. 20-21.   15.

Ministère de la Marine. Mines Sous-Marines Automatiques Harlé à Antennes Type Z; Harlé et Cie., Paris. 27 p.   16.

1. *Min Pac Goes Deep Sea Hunting;* All Hands, No. 546, Jul. 1962; Washington, D.C. pp. 2-4.

2. Mintzer, Leonard Murney; *For the Defense;* U.S. Naval Institute Proceedings, Vol. LXX, No. 11, Nov. 1944; Annapolis, Maryland. pp. 1363-1371.

3. Mitchell, Donald W.; *History of the Modern American Navy;* Knopf, 1946; New York.

4. Mitchell, Donald W.; *Russian Mine Warfare: The Historical Record;* Journal of the Royal United Service Institution, Vol. CIX, No. 633, Feb. 1964; London. pp. 32-39.

5. Mitchell, Mairin; *The Maritime History of Russia;* Sidgwick and Jackson, 1948; London.

6. *Moderne Søminer;* Tidsskrift for Sovaesen, Vol. XCVIII, 1927; Copenhagen. pg. 537.

7. Moerath, J. N.; *On the Protection of Vessels Against Torpedoes;* Engineer, Vol. XXXIII, Apr. 5, 1872; London. pg. 245.

8. Moerath, J. N.; *On the Protection of Vessels Against Torpedoes;* Journal of the Royal Society of Arts, Vol. XX, 1872; London. pp. 362-363.

9. Moerath, J. N.; *On the Protection of Vessels Against Torpedoes;* Transactions of the Institute of Naval Architects, Vol. XIII, 1872; London.

10. *The "Monitor" Torpedo;* Scientific American, Vol. IX, No. 4, Jul. 25, 1863; New York. pg. 56.

11. Monosterev, N.; *La Guerra de Minas in el Baltica 1914-1915;* Ministro de Marina, Rivista General de Marina, Vol. CXVIII, 1936; Madrid. pp. 327-333.

12. *Monster Rally;* Time, Vol. LVI, No. 21, Nov. 20, 1950; New York. pg. 30.

13. Montgery; *Mémoire sur les Mines Flottantes et les Pétards Flottans où Machines Infernales Maritimes;* Boahelier, Libraire pour la Marine, 1819.

14. *A Month or So From Now, The Mining Will Begin to Bite;* U.S. News and World Report, Vol. LXXII, No. 24, Jun. 12, 1972; Washington, D.C. pp. 35-36.

The Months News; All Hands, No. 317, Aug. 1943; Washington, D.C. pg. 29. 1.

Morison, Samuel Eliot; *History of United States Naval Operations in World War II, Vol. XI: The Invasion of France and Germany 1944-1945;* Little, Brown and Co., 1957; Boston. 2.

Morison, Samuel Eliot; *History of United States Naval Operations in World War II, Vol. XV: Supplement and General Index;* Little, Brown and Co., 1957; Boston. 3.

Morison, Samuel Eliot; *The Oxford History of the American People;* Oxford University Press, 1965; New York. 4.

Morison, Samuel Eliot; *The Two-Ocean War;* Little, Brown and Co., 1963; Boston. 5.

Motor Minesweeping Flotilla; British Motor Ship, Vol. XXVI, No. 307, Aug. 1945. pp. 164-167. 6.

Mouillage de Mines en Mer du Nord; L'Illustration, Vol. CCV, No. 5068, Apr. 20, 1940; Paris. pg. 388. 7.

Moving in Mine-Sown Waters; Journal of the U.S. Artillery, Vol. XLII, Jul-Aug. 1914; Ft. Monroe, Virginia. pp. 375-376. 8.

Moving in Mine Sown Waters; United Service Gazette, Vol. CLXIII, Sept. 10, 1914; London. pp. 205-206. 9.

Muir, H. J.; *Aerial Mines;* New Statesman and Nation, Vol. XVII, No. 426, Apr. 22, 1939; London. pp. 608-609. 10.

Murder at Sea and the Blockade of German Experts; Spector, Vol. CLXIII, Nov. 24, 1939; London. pg. 729. 11.

(Murray), Army and Navy Journal, Vol. XLII, No. 17, Dec. 24, 1904; New York. pg. 430. 12.

Mysterious Mining of British Destroyers Off the Albanian Coast; Illustrated London News, Vol. CCIX, No. 9, 1946; London. pp. 509, 511. 13.

Mystery of the Magnetic Mine; U.S. Naval Institute Proceedings, Vol. LXVI, May 1940; Annapolis, Maryland. pp. 754-756. 14.

1. *NATO Glossary of Mine Warfare Terms;* NATO Military Agency for Standard-
      izations.

2. *Nauticus;* Jahrbuch für Deutchlands Seeinteressen Sechzelmter Jahrgang,
      Mittler & Sohn, Vol. XVI, 1914; Berlin.

3. *Naval Mines;* Colliers Encyclopedia, Vol. XVII, Crowell-Collier Publishing
      Co., 1965. pp. 231-235.

4. *Naval Mining and Degaussing;* Her Majesty's Stationery Office, 1946; London.

5. *Naval Mining and Degaussing Exhibition;* Engineering, Vol. CLXI, Aug. 9,
      1946; London. pg. 129.

6. *Naval Operations Analysis;* U.S. Naval Institute, 1968; Annapolis, Maryland.

7. *Naval Weapons and Their Uses;* U.S. Naval Institute, 1943; Annapolis, Maryland.

8. *Les Navires Mouilleurs de Mines Sous-Marines "Pluton" et "Cerbere";* Le
      Génie Civil, Vol. LXII, No. 26, Apr. 26, 1913; Paris. pp. 501-503.

9. *Navy Department, Office of Naval Intelligence Historical Section Pub. #7,
      The American Naval Planning Section London;* U.S. Government Printing
      Office, 1923; Washington, D.C.

10. *Navy is Ready to Clear Mines from Harbors of North Vietnam;* U.S. Naval
      Institute Proceedings, Vol. XCIX, No. 3, Mar. 1973; Annapolis,
      Maryland. pp. 122-123.

11. *Navy Makes 1,000 Mines Per Day;* U.S. Naval Institute Proceedings, Vol. XLIV,
      No. 11, Nov. 1918; Annapolis, Maryland. pg. 2620.

12. *Navy Ordnance Activities World War 1917-1918;* U.S. Government Printing
      Office, 1920; Washington, D.C.

13. *Navy Ordnance Big Mine Production;* U.S. Naval Institute Proceedings, Vol.
      XLIV, No. 10, Oct. 1918; Annapolis, Maryland. pp. 2382-2383.

14. *Navy Training Helpers: Sea Lions to Clear Mines;* Evening Star and Daily
      News, Vol. CXX, No. 284, Tuesday, Oct. 10, 1972; Washington, D.C.
      pg. A-10.

15. *Navy Weighs Scrapping of "Mined" Ship;* Evening Star, Vol. CXX, No. 346,
      Monday, Dec. 11, 1972; Washington, D.C. pg. A-8.

Navy's Newest Minesweeper (Affray) Commissioned; Army-Navy-Air   1.
    Force Register, Vol. LXXX, No. 4124, Dec. 20, 1958; Washington,
    D.C. pg. 8.

La Neuville, C.; *Mines et Dragage des Mines;* Journal de la Marine   2.
    Merchande, Vol. XXII, No. 1083, Jan. 4, 1940; Paris. pp. 8-10.

Newbolt, Henry; *History of the Great War - Naval Operations;* Longmans,   3.
    Green, Vol. V, 1931; London.

New Developments in Hull Design of Wooden Ships; American Society of   4.
    Naval Engineers Journal, Vol. LXVI, No. 4, Nov. 1954; Washington,
    D.C. pp. 867-878.

Newell, John S.; *Lecture Notes on Torpedoes, Clearing Channels and the*   5.
    Use of Explosives in the Removal of Obstructors; (incomplete title),
    Rockwell and Churchill, 1886; Boston.

Newest Navy Minesweeper "Assurance" Commissioned; Army-Navy-Air   6.
    Force Register, Vol. LXXIX, No. 4122, Dec. 6, 1958; Washington,
    D.C. pg. 12.

Newman, Al; *Minelaying - Bit of Bind;* Newsweek, Vol. XXI, No. 9,   7.
    Mar. 1, 1943; Washington, D.C. pg. 23.

Newman, James Ray; *The Tools of War;* Doubleday, Doran & Co., Inc.,   8.
    1942; Garden City, New York.

New Methods of Planting Torpedoes; Scientific American, Vol. X, No. 23,   9.
    Jun. 4, 1864; New York. pg. 362.

New Minesweeper Built by General Engineering; Log, Vol. XXXIX, No. 7,   10.
    Jul. 1944; San Francisco. pp. 66-67, 70.

A New Rebel Torpedo; Scientific American, Vol. IX, No. 25, Dec. 19, 1863.   11.
    pg. 388.

New Rebel Torpedo; Scientific American, Vol. XI, No. 15, Oct. 8, 1864;   12.
    New York. pg. 228.

New Role for Submarine Mines; Worlds Work, Vol. XXIV, Nov. 1914;   13.
    Garden City, New York. pp. 92-95.

New Torpedo; The Engineer, Vol. XVIII. Oct. 21, 1864; London. pg. 254.   14.

1. *A New Torpedo;* Engineer, Vol. XIX, 1865; London. pg. 253.

2. *A New Torpedo;* Engineer, Vol. XX, 1866; London. pg. 422.

3. *The New Torpedo in Charleston Harbor;* Scientific American; Vol. IX,
        No. 11, Sept. 12, 1863; New York. pg. 164.

4. *New Type Minesweeper, MSB-5;* Army and Navy Journal, Vol. XC, No. 13,
        Nov. 29, 1952; Washington, D.C. pg. 398.

5. *New Wooden Ships Join "Sweep Fleet";* All Hands, No. 444, Feb. 1954;
        Washington, D.C. pp. 10-13.

6. Nikolaev, Valerii Vasilevich; *Morskie Sapery;* 1967; Moscow.

7. Nikolayev, B.; *Mine and Torpedo Weapon Systems;* Starshina-Serzhant,
        No. 11, Nov. 1970.

8. *Nini Especial Empleada por los Submarinas Minadores Alemanes del Tipo
        U.C.;* Revista General de Machiva, Afia 45, 1922; Madrid.
        pp. 493-508.

9. *Nixon on Vietnam: We Will Return to Negotiation Table;* U.S. News and
        World Report, Vol. LXXIII, No. 2, Jul. 10, 1972; Washington, D.C.
        pp. 51-52.

10. Noalhat, H.; *Les Torpilles et les Mines Sous-Marines;* Berger-Levrault
        & Cie., 1905; Paris.

11. *Noch einmal der Minenkrieg;* Ueberall, Vol. IX, No. 23, 1906; Berlin.
        pp. 23-24.

12. Nock, C. F.; *Mines Ahead;* Mines and the Royal Naval Minewatching
        Service, Shipbuilding, Vol. LXXIX, Nov. 13, 1952; London. pg. 633.

13. Noel, Cdr G. H. U.; *The Gun Ram, and Torpedo;* Manoeuvres and tactics
        of a naval battle in the present day; Griffin, 1889; Portsmouth.
        354 p.

14. *Nogle Bemaerkninger om U-Baade og passive Scminer;* Tidsskrift for
        Sovaesen, Vol. LXXIX, 1908; Copenhagen. pg. 357.

15. *Nogle Betragning Over Mineskringssporgsmaalet;* Tidsskrift for Sovaesen,
        Vol. XCVIII, 1927; Copenhagen. pg. 177.

Nolan, E. H.; *Illustrated History of the War Against Russia;* James S.   1.
    Virtue, 2 volumes, 1856; London.

Non-Magnetic Gear Beats Red Mine; Iron Age, Vol. CLXXII, Dec. 24, 1953; 2.
    Philadelphia, Pennsylvania. pg. 29.

Nonmagnetic Minesweepers (for Belgium); Military Review, Vol. XXXVII,   3.
    No. 7, Nov. 1957; Ft. Leavenworth. pg. 71.

Normand, J. A.; *Notre Puissance Navale;* Berger-Levrault & Cie., 1900; Paris. 4.

The Northern Barrage; Mine Force, U.S. Atlantic Fleet; Edited by All Hands, 5.
    U.S. Naval Institute, 1919; Annapolis, Maryland.

The Northern Barrage and Other Mining Activities; Publication No. 2, Navy 6.
    Department, Office of Naval Records and Library Historical Section,
    U.S. Government Printing Office, 1920; Washington, D.C.

The Northern Barrage (Taking Up the Mines); Publication No. 4, Navy Depart- 7.
    ment, Office of Naval Records and Library Historical Section, U.S.
    Government Printing Office, 1920; Washington, D.C.

Northern Theatre: Spring Offensive; Time, Vol. XXXV, No. 16, Apr. 15, 1940; 8.
    New York. pg. 25.

North Sea Danger Area; Naval News, United Service Gazette, Vol. CLXIII, 9.
    Nov. 5, 1914; London.

North Sea Mine Sweepers Waging Final Sweep; U.S. Naval Institute Pro- 10.
    ceedings, Vol. XLV, No. 6, Jun. 1919; Annapolis, Maryland. pg. 1029.

North Sea Submarine Barrage; Current History Magazine, New York Times, 11.
    Vol. IX, Part I, Dec. 1918; New York. pp. 489-491.

Notes on Aids to Submarine Hunting; Airship Department of the Admiralty, 12.
    C.B. No. 01454, Apr. 1918; London.

Notes on Torpedoes; Engineering, Vol. XXII, No. 10, Jul. 28, 1876; London. 13.
    pp. 67-68.

Notes on Torpedoes; Engineering, Vol. XXII, No. 13, Dec. 1, 1876; London. 14.
    pp. 459-460.

Notre Marine de Guerre en 1899. Les Vices de Son Organisation. Un Programme 15.
    de Réformes; Berger-Levrault & Cie., 1899; Paris.

Notre Marine de Guerre. Réformes Essentielles; Berger-Levrault & Cie., 1904; 16.
    Paris.

1. *Novel Use of the Camera Obscura;* Van Nostrand's Engineering Magazine, Vol. I, No. 3, Mar. 1869; New York. pg. 288.

2. Nutting, William Washburn; *The Mysterious Paravane;* International Marine Engineer, Vol. XXIV, No. 4, Apr. 1919; New York. pp. 289-293.

3. *Oberon Experiments;* Engineer, Vol. XXXVIII, Aug. 28, 1874; London. pg. 176.

4. *Oberon Experiments;* Engineer, Vol. XXXVIII, Sept. 4, 1874; London. pg. 185.

5. *Oberon Experiments;* Engineer, Vol. XXXVIII, Sept. 25, 1874; London. pg. 232.

6. *The Oberon Experiments;* The Engineer, Vol. XLII, Jul. 7, 1877; London. pp. 14-15.

7. *The Oberon Submarine Mine Experiments;* Army and Navy Journal, Vol. XII, No. 45, Jun. 19, 1875; New York. pp. 715-716.

8. *The Oberon Submarine Mine Experiments;* The Engineer, Vol. XXXIX, May 28, 1875; London. pg. 361.

9. *The Obry Apparatus;* Journal of the U.S. Artillery, Vol. XXIV, No. 2, Sept-Oct. 1905; Ft. Monroe, Virginia. pg. 179.

10. *Observation Mines for Harbor Protection;* Scientific American, Vol. CXIII, No. 20, Nov. 13, 1915; New York. pg. 430.

11. *The Offensive Mine Laying Campaign Against Japan;* U.S. Strategic Bombing Survey, Naval Analysis Division, Nov. 1, 1946.

12. Office of the Chief of Staff of the Royal Italian Navy; *The Italian Navy in the World War 1915-1918;* Provveditorato Generale Dello Stato, May 1927.

13. *Official History (Naval and Military) of the Russo-Japanese War;* Historical Section of the Committee of Imperial Defence, Vol. II, 1912.

14. *Official Records of the Union and Confederate Navies in the War of the Rebellion;* U.S. Government Printing Office, 31 volumes, 1894-1927; Washington, D.C.

15. O'Hearn, Edward P.; *Explosives;* Smithsonian Institution Annual Report 1914; U.S. Government Printing Office, 1915; Washington, D.C. pp. 269-271.

16. *Oil-Engined Minesweepers;* Marine Engineer; Vol. LXIX, No. 823, Feb. 1946; London. pp. 76-80.

Oil-Engined Minesweepers; Shipbuilder, Vol. LIII, No. 444, Mar. 1946; 1.
    London. pp. 120-122.

Oliver, F. L.; *Inventions Watch on the Sea;* Christian Science Monitor 2.
    Magazine, May 18, 1940; Boston. pg. 6.

Om Anvendelsen af forankrede Miner (Prisafhandling tilkendt SLS 3.
    Guldmedaille); Tidsskrift for Sovaesen, Vol. XCIII, 1922; Copenhagen.

Om Anvedelsen af Passive Sominer i den Kinesiske-Japanske Krig 1894 og i 4.
    den Spansk-Amerikanse Krig 1898; Tidsskrift for Sovaesen, Vol. LXX,
    1899; Copenhagen. pg. 457.

Om Minesogning og Minerydning; Tidsskrift for Sovaesen, Vol. C, 1929; 5.
    Copenhagen. pp. 434, 481, 543.

Om Minevaesenets Standpunkt før og Sominens Anvendelse under Verdenskrigen 6.
    1914; Tidsskrift for Sovaesen, Vol. LXXXVI, 1915; Copenhagen. pg. 49.

OP 342; *Naval Defense Mine Mark II;* U.S. Government Printing Office, 1909; 7.
    Washington, D.C.

OP 344; *U.S. Naval Defense Mines Mark III, Mod 1 and Mark IV;* U.S. Government 8.
    Printing Office, May 1916; Washington, D.C.

OP 344A; *U.S. Naval Defense Mine Mark IV, Mod 2;* U.S. Government Printing 9.
    Office, Dec. 1917; Washington, D.C.

Opankring af Selvvirken de Miner; Tidsskrift for Sovaesen, Vol. CII, 1931; 10.
    Copenhagen. pg. 267.

Operation Minesweep; International Nickel Company, Vol. XXVII, No. 1, Jun. 11.
    1957; New York. pp. 30-33.

OPNAV NOTICE 5430; From: Chief of Naval Operations: To: Distribution List; 12.
    Subject: Change to the Organization of the Office of the Chief of
    Naval Operations; OPNAVNOTE 5430; OP-09B; Ser: 22980P09B; Nov. 27,
    1972.

Oppenheim, Lassa F. L.; *International Law;* H. Lauterpacht, Longmans, Green, 13.
    7th Edition, 1952; London. pp. 681-682.

Ordnance and Gunnery; U.S. Naval Institute, 1910; Annapolis, Maryland. 14.

Ordnance Notes; Ordnance Office, U.S. War Department, No. 207. 15.

1. *Organization of a Torpedo Corps;* Army and Navy Journal, Vol. VI, No. 50, Jul. 31, 1869; New York. pg. 788.

2. Ormsbey, Eugene; *Getting Rid of the Boom is Disposaleers Business;* All Hands, No. 494, Mar. 1958; Washington, D.C. pg. 25.

3. Orvin, Maxwell Clayton; *In South Carolina Waters, 1861-1865;* Southern Printing and Publishing Company, 1961; Charleston, South Carolina.

4. *Pacific Cinderella;* All Hands, No. 351, U.S. Navy Bureau of Personnel, Jun. 1946; Washington, D.C. pp. 21-27.

5. Padgett, Harry E.; *Newest Minesweepers in the U.S. Navy;* U.S. Naval Institute Proceedings, Vol. LXXXIX, No. 7, Jul. 1963; Annapolis, Maryland. pp. 172-173.

6. Padgett, Harry E.; *Reviews (America's Use of Sea Mines by R. C. Duncan);* U.S. Naval Institute Proceedings, Vol. XC, No. 4, Apr. 1964; Annapolis, Maryland. pp. 120-122.

7. Palmer, Francis Ingraham; *Descriptive Account of an Horizontal Acting and Disconnecting Spar Torpedo-Ram Fitted with Auxiliary Needle.* 73 p.

8. Palmer, Wayne F.; *Submarine Mining Orphan Child of the Service;* U.S. Naval Institute Proceedings, Vol. LX, No. 11, Nov. 1934; Annapolis, Maryland. pg. 1582.

9. Pan-Se-Tcheng; *Description du Tonnere Sous-Marine.* This is probably the most interesting document in the whole collection. It appears to describe (in Chinese block prints) a pressure-operated mine. The date is 1843. The title page states "Treasured at the Studio of Sea Coral Immortal."

10. *Parade of Navy Mine Sweeping Force;* Army and Navy Journal, Vol. LVII, No. 9, Nov. 1, 1919; New York. pg. 276.

11. *Paravane, It Foiled the German Mines;* Literary Digest, Vol. LXI, No. 7, May 17, 1919; New York. pp. 23-24.

12. Parsons, William Barclay; *Robert Fulton and the Submarine;* Columbia University Press, 1922.

13. Patiens; *La Défense Nationale et la Défense des Côtes;* Berger-Levrault & Cie., 1894; Paris.

Patrols and Sweepers; The Navy, Vol. XXI, No. 5, May 1916; London.   1.
        pp. 128-130.

Patterson, Andrew, Jr.; *Mining, A Naval Strategy;* Naval War College   2.
        Review, Vol. XXIII, No. 9, May 1971; Newport, Rhode Island.
        pp. 52-66.

Paullin, Charles Oscar; *Commodore John Rodgers;* A. H. Clark Co., 1910;   3.
        Cleveland. pp. 204-207.

Pavlovich, N. B.; *Flot y Pervoi Mirovoi Uione;* (The Fleet in the First   4.
        World War), 1964; Moscow.

Pawle, Gerald; *The Secret War;* U.S. Naval Institute Proceedings, Vol.   5.
        LXXXIV, No. 6, Jun. 1958; Annapolis, Maryland. pg. 116.

Pawle, Gerald; *The Secret War 1939-1945;* William Sloane Associates, Inc.,   6.
        1957; New York.

"P.C."; *Les Mines Sous-Marines pour la Défénse de Ports;* Le Génie Civil,   7.
        Vol. LXV, No. 19, Sept. 5, 1914; Paris. pp. 360-362.

Pelissier, Jean; *Cap sur Hammerfest;* Robert Laffont, 1962; Paris. 164 p.   8.
        [Roman sur la peche à la morue.]

Perlia, Sigmund Naumovich; *Udar pod Vodol;* 1945; Moscow.   9.

Perry, Milton F.; *Infernal Machines;* Louisiana State University Press,   10.
        1965.

Pesce, G. L.; *La Navigation Sous-Marine;* Vuibert and Nony, 1906; Paris.   11.
        pp. 476-480.

Pfankuchen, Llewellyn; *A Documentary Textbook in International Law;*   12.
        Rinehart & Company, 1940; New York.

Philipson, Coleman; *International Law and the Great War;* Dutton, 1916;   13.
        New York. pg. 372.

Piehler, D. and N. Hirshberg; *Display Systems State of the Art;*   14.
        U440-72-081; Electric Boat Division, General Dynamics Corporation,
        Aug. 1972.

Piffera y Gallindo, Juan de la; *Instalacion del Barrje de Minas del Mar*   15.
        del Norte; Revista General de Marina, 1922; Barcelona. pp. 207-232.

1. Piron, Cap. du Génie F. P. S.; Études sur les Canonnières Cuirassées.

2. Piterskii, Nikolai A. (Editor); Combat Path of the Soviet Navy; Military
    Publishing House, 1966; Moscow. pg. 20.

3. "P.L."; Mines Sous-Marines et Mouillers de Mines; Journal de Marine, le
    Yacht, Vol. XXXVII, No. 1888, May 16, 1914; Paris. pp. 305-306.

4. "P.L."; Mines Sous-Marines et Mouillers de Mines; Journal de Marine, le
    Yacht, Vol. XXXVII, No. 1889, May 23, 1914; Paris. pg. 322.

5. (Planters, Mine) Army and Navy Journal, Vol. XLVI, No. 10, Nov. 7, 1908;
    New York. pg. 252.

6. Playing With Death in the Dardanelles; Technical World Magazine, Vol. XXIII,
    No. 5, Jul. 1915; New York. pp. 580-581.

7. Pluddemann, Martin; Comments of Rear Admiral Pluddemann, German Navy on
    the Main Features of the War with Spain; Office of Intelligence, U.S.
    Government Printing Office, 1898; Washington, D.C. [War Notes No. 2.]

8. Pluddemann, Martin; Main Features of the Spanish American War; U.S. Naval
    Institute Proceedings, Vol. XXIV, No. 4, Dec. 1898; Annapolis,
    Maryland. pp. 771-788.

9. Pluddemann, Martin; Modernes Seekriegswesen; Mittler und Sohn, 1902; Berlin.
    pp. 988-991.

10. Politovsky, E. S.; From Libau to Tsushima; E. P. Dutton and Co., 1906;
    New York. [Translated by F. R. Godfrey.]

11. Polluted Water; Time, Vol. XLV, No. 21, May 21, 1945; New York. pg. 29.

12. Polmar, Norman; The Mine as a Tool of Limited War; U.S. Naval Institute Pro-
    ceedings, Vol. XCIII, No. 6, Jun. 1967; Annapolis, Maryland. pp. 103-105.

13. Port Arthur, 1904 the Results at; Sea Power, Dec. 12, 1911; New York.

14. Porter, D. C.; The Naval History of the Civil War; Sherman, 1886; New York.
    xvi, 843 p.

15. Les Porteurs de Mines "Cerbere et Pluton"; Journal de Marine, le Yacht,
    Vol. XXXVI, No. 1868, Dec. 27, 1913; Paris. pp. 837-838.

Portlock, Ronald; *Underwater Warfare;* The Navy, Vol. LXX, No. 8, Aug. 1965;   1.
London. pg. 272.

Ports of the World; Benn Brothers, Ltd., 21st Edition, 1967; London.   2.

Postan, M. M., D. Hay and J. D. Scott; *History of the Second World War*   3.
*Design and Development of Weapons - Studies in Government and Indus-
trial Organization;* Her Majesty's Stationery Office, 1964; London.

Potter, E. B. (Editor); *The United States and World Sea Power;* Prentice-   4.
Hall, Inc., 1955; Englewood Cliffs, New Jersey.

Potter, E. B. and C. W. Nimitz (Editors); *Sea Power: A Naval History;*   5.
Prentice-Hall, 1960; Englewood Cliffs, New Jersey.

Potter, E. B. and C. W. Nimitz (Editors); *Triumph in the Pacific;* Prentice-   6.
Hall, 1960; Englewood Cliffs, New Jersey.

Potter, Pitman B.; *The Freedom of the Seas in History, Law and Politics;*   7.
Longmans, Green & Co., 1924; New York.

Powers, Robert D., Jr.; *Blockade: For Winning Without Killing;* U.S. Naval   8.
Institute Proceedings, Vol. LXXXIV, No. 8, Aug. 1958; Annapolis,
Maryland. pp. 61-66.

Powers, Robert D., Jr.; *International Law and Open Ocean Mining;* JAG   9.
Journal, Vol. XV, No. 4, Jun. 1961; Washington, D.C. pp. 55-58, 71.

Pratt, Fletcher; *Civil War on Western Waters;* Holt, 1956; New York.   10.

Pratt, William V.; *The Mine Continues to Take a Deadly Toll;* Newsweek,   11.
Vol. XX, No. 14, Oct. 1942; Washington, D.C. pg. 26.

Pratt, William V.; *The Mine as a Weapon Against Japan;* Newsweek Magazine,   12.
Vol. XX, No. 26, Dec. 28, 1942; New York. pg. 21.

Pressing and Welding Submarine Mines; Electrical Review, Vol. CXXIII, No.   13.
3424, Jul. 9, 1943; London. pp. 35-39.

Pressure on North Vietnam: Enough to Bring Peace?; U.S. News and World   14.
Report, Vol. LXXII, No. 26, Jun. 26, 1972; Washington, D.C. pp. 29-30.

Professional Notes; Journal of the U.S. Artillery, Vol. XXI, No. 2, Mar-Apr.   15.
1904; Ft. Monroe, Virginia. pg. 215.

1. *Professional Papers No. XI, U.S. Engineers School;* Sgt. Carmichiel and Pvt. Beck, Printers, 1888; Willets Point, New York Harbor.

2. *Professional Papers of the Corps of Royal Engineers. 1892, No. 4, Confidential Series. On the Principles of Submarine Mining Defence, and its Connection with Coast Batteries;* Major R. M. Ruck, RE. W. and J. Mackay, 1892; Chatham.

3. *Professional Papers on Subjects Connected with the Duties of the Royal Engineers;* John Weale, Vol. VII, 1845; London. [Paper III.]

4. *Professional Papers on Subjects Connected with the Duties of the Corps of Engineers Contributed by Officers of the Royal Engineers, New Series;* Jackson and Son, Vol. XVII, 1869; Woolwich.

5. *Professional Papers on Subjects Connected with the Duties of the Corps of Royal Engineers Contributed by Officers of the Royal Engineers;* Vol. XV, 1866. [Paper I.]

6. *Program for Uddannelsen of Mandkobet ved late Ingenieurballions;* Danish War Ministry, 1870; Copenhagen.

7. *La Protection des Navires Marchands contre les Mines;* Livre à l'usage des capitaines de navires munis d'appareils Otter, Societé des Chantiers et Ateliers Augustin Norman, 1917; Le Havre. 17 p. Secret et confidentiel No. 226. [It is no longer secret and confidential.]

8. *The Protection of Merchant Vessels Against Moored Mines;* Vickers Ltd., 1917; London.

9. *Protection of Ships Against Mines;* Engineer, Vol. CXXVII, No. 3297, Mar. 7, 1919; London. pp. 222-224.

10. *Protection of Ships Against Mines;* Engineer, Vol. CXXVII, No. 3300, Mar. 28, 1919; London. pp. 293-294.

11. Puleston, W. E.; *The Dardanelles Expedition;* U.S. Naval Institute, 1926; Annapolis, Maryland.

12. Pyke, G.; *Reply to Muir, H. J. Aerial Mines of April 22, 1939;* New Statesman & Nation, Vol. XVII, No. 428, May 6, 1936; London. pg. 685.

13. "Quarterdeck"; *Future Naval Warfare;* Sea Power, Vol. IX, No. 6, Dec. 1920; Washington, D.C. pp. 281-282.

14. *Radio-Controlled Drone Boats Used in Vietnam Minesweeping;* U.S. Naval Institute Proceedings, Vol. XCVI, No. 2, Feb. 1970; Annapolis, Maryland. pp. 123-124.

RADM Dietrich Leads Giant Mine Maneuver (Operation Lurk Deep); Army and Navy   1.
    Journal, Vol. XCIV, No. 9, Nov. 3, 1956; Washington, D.C. pg. 18.

Raids on the Ruhr; Newsweek, Vol. XXI, No. 19, May 10, 1943; Washington,   2.
    D.C. pg. 24.

Rairden, Percy Wallace, Jr.; *The Importance of Mine Warfare;* U.S. Naval In-   3.
    stite Proceedings, Vol. LXXVIII, No. 8, Aug. 1952; Annapolis, Maryla..d.
    pp. 847-849.

Rairden, Percy Wallace, Jr.; *The Junior Officer in Mine Warfare;* U.S. Naval   4.
    Institute Proceedings, Vol. LXXIX, No. 9, Sept. 1953; Annapolis,
    Maryland. pp. 977-979.

Ranson, M. A.; *Little Gray Ships;* U.S. Naval Institute Proceedings, Vol.   5.
    LXII, No. 9, Sept. 1936; Annapolis, Maryland. pp. 1280-1294.

Rapport de la Commission des Défense Sous-Marines; Texte et dessins, Litho   6.
    de Ministère de la Marine, 1866-1867; Paris.

Rapport de la Commission des Mines Automatiques Sous-Marines (9 Dec 1911);   7.
    Imprimerie Nationale, 1912; Paris. 114 p.

Rapport de la Haute Commission Militaire; Exposition Universelle de 1867,   8.
    Librairie Administrative de Paul Dupont, 1869; Paris.

The R.C.A.F. Overseas the First Four Years; Oxford University Press, 1944;   9.
    Toronto.

Rear Admiral Joseph Strauss Arrives Home; Army and Navy Journal, Vol. LVII,   10.
    No. 8, Oct. 25, 1919; Washington, D.C. pg. 233.

A Rebel Infernal Machine; Scientific American, Vol. VIII, No. 2, Jan. 10,   11.
    1863; New York. pg. 19.

A Rebel Torpedo; Frank Leslie's Illustrated Newspaper, Vol. XIII, 1861;   12.
    New York. pg. 352.

The Rebel Torpedoes; Scientific American, Vol. X, No. 25, Jun. 18, 1864;   13.
    New York. pg. 390.

Receive Minesweepers (Denmark); Military Review, Vol. XXXVI, No. 12, Mar.   14.
    1957; Ft. Leavenworth. pg. 72.

Recent Improvements in the Wire Drag Used by the U.S. Coast and Geodetic   15.
    Survey: Its Possible Use as a War Machine; Engineering News, Vol.
    LXXII, No. 9, Aug. 27, 1914; New York. pp. 446-447.

1. *The Reckless Strewing of Mines;* Marine Engineer, Vol. XXXVII, Part V,
        No. 447, Dec. 1914; London. pg. 131.

2. *Recognition;* Time, Vol. XXXV, No. 1, Jan. 1, 1940; New York. pg. 28.

3. Record of proceedings of a court of inquiry convened to inquire into the
        loss of the U.S.B.S. Maine in the harbor of Havana, Cuba on the night
        of February 15, 1898; 55th Congress, 2nd Session, S. Doc. 207, U.S.
        Government Printing Office, 1898; Washington, D.C.

4. *Records of Conference for Limitations of Armaments;* Jun. 20 - Aug. 4, 1927;
        Geneva. pp. 74, 79, 80, 144.

5. *Recueil de Mémoirs sur la Démagnetization lvs (?) Devant l'Institution
        des Ingenieurs Électriciens Britanniques;* Apr. 4-5, 1946; London.

6. Reed, David; *Mission: Mine Haiphong;* Reader's Digest, Vol. CII,
        No. 610, Feb. 1973; Pleasantville, New York. pp. 76-81.

7. *Regulations for Mine Planters;* U.S. War Department, U.S. Army, U.S. Government
        Printing Office, 1907; Washington, D.C.

8. Reigart, J. Franklin; *The Life of Robert Fulton;* C. G. Henderson Co.,
        1856; Philadelphia.

9. *Removing Mines in North Sea;* Army and Navy Journal, Vol. LVI, No. 33,
        Apr. 19, 1919; New York. pg. 1163.

10. *Repeating Fire Arms;* United States Magazine; Vol. IV, No. 3, Mar. 1857;
        J. M. Emerson & Co., New York. pp. 221-247.

11. *Report of Inspector of Submarine Defences for 1899;* Eyre and Spottiswoode,
        1900; London.

12. *Report of Loss of Two British Vessels;* Fairplay, Vol. XLIX, Aug. 15, 1907;
        London. pg. 244.

13. *Report of the Experiments Conducted by the War Office Torpedo Committee
        at Stokes Bay and Elsewhere During the Year 1875-6;* War Office, 1876.

14. *Report of the Secretary of the Navy;* Annual Reports, U.S. Government Printing
        Office, 1900 and sequentially; Washington, D.C.

15. *Report of Torpedo Committee Appointed 29th July 1870 by HRH the Field-
        Marshal Commanding in Chief, 'For the Purpose of Deciding on the
        Form, Composition, and Machinery of Torpedoes';* HMSO, Harrison
        and Sons, 1875; London.

Reuter, Herbert C.; *History of Submarine Mines 1585-1920;* Submarine Mine   1.
    Depot, 1939; Ft. Monroe, Virginia.

Reuter, Herbert C.; *A Summary of Historical Information Pertaining to Con-* 2.
    trolled Submarine Mining; U.S. Submarine Mine Depot, 1949; Ft. Monroe,
    Virginia.

Reventlow, E. Graf; *Minenkrieg;* Ueberall, Vol. X, 1907; Berlin. pp. 162-167. 3.

Review of P. D. Bunker Article; Army and Navy Journal, Vol. LI, No. 34,   4.
    Apr. 25, 1914; New York. pg. 1064.

Review of U.S. Minesweepers; Army and Navy Journal, Vol. LVII, No. 13,   5.
    Nov. 29, 1919; New York. pp. 395-396.

Revolutionary Mine Detector Equipment; Engineer, Vol. CCXII, Oct. 27, 1961; 6.
    London. pg. 712.

Reymond, P.; *Le Sous-Marin Mouiller de Mines;* Moniteur de la Flotte, Vol. 7.
    LX, No. 1, Jan. 4, 1913; Paris.

"R.G.E."; *Der Minenkrieg in Ostasien;* Ueberall, Vol. VI, No. 26, 1904; 8.
    Berlin. pp. 485-486.

Richardson, J. B.; *Coast Defense Against Torpedo-Boat Attack;* Journal of 9.
    the Military Service Institution, Vol. XXIV, 1899; Governor's Island,
    New York. pp. 259-275.

Richardson, J. B.; *Coast Defense Against Torpedo Boat Attack;* Journal of 10.
    the U.S. Artillery, Vol. XI, No. 1, Jan-Feb. 1899; Ft. Monroe,
    Virginia. pp. 65-83.

Riggs, Jerry; *U.S. Mine Sweeping Boats Keep Clear the River Paths to Saigon;* 11.
    Navy, Vol. X, No. 5, May 1967; Washington, D.C. pp. 15-18.

Riley, David R.; *Mine Warfare in World War Two;* U.S. Naval Postgraduate 12.
    School, Operations Analysis Course, 1972; Monterey, California.

(Rio de Janerio); Nautical Magazine, Dec. 1866; London. pp. 683-684. 13.

Robert Fulton's Torpedoes; Scientific American, Vol. LXXVIII, No. 23, 14.
    Jun. 4, 1898. pg. 361.

Robertson, W. W.; *U-Boat Activities 1917-1918;* U.S. Naval Institute Pro- 15.
    ceedings, Vol. LXVIII, No. 5, May 1942; Annapolis, Maryland. pp. 672-
    676.

1. Robinson, Charles Napier; *Royal Navy Handbooks, Torpedoes and Torpedo Vessels;* Robinson Editor, by Lieut. G. E. Armstrong, George Bell, 1896; London. [Second Edition, 1901]

2. Robinson, Reed A.; *Degaussing-Magnetic "Irresistibility";* Sperry Scope, Vol. XIV, No. 9, 1958; Brooklyn, New York. pp. 16-19.

3. Robinson, S. S.; *A History of Naval Tactics from 1530 to 1930;* U.S. Naval Institute, 1942; Annapolis, Maryland.

4. Rocholl, Erich; *Die Frage der Minen im Seekrieg;* A. Ebel, 1910; Marburg.

5. Rocholl, Erich; *Die Frage der Minen im Seekrieg;* R. Noske Borora-Leipzig, 1910.

6. Roden, Ernest K.; *Submarine Mines;* The Mechanic Arts Magazine, Vol. IV, No. 3, Whole No. 39, Apr. 1899; Scranton. pp. 118-122.

7. Roebling Manufacturing; *Wire-Roping the German Submarine;* John A. Roebling's Sons Co., 1920; Trenton, New Jersey.

8. Rodgers, Hamp and Ron Gorman; *Airdales and Black Shoes Sweep Mines;* Our Navy Magazine, Vol. LXVII, No. 3, Mar. 1972; Brooklyn, New York.

9. Rodgers, Robert H.; *America's Best Weapon;* U.S. Naval Institute Proceedings, Vol. XCI, No. 9, Sept. 1965; Annapolis, Maryland. pp. 106-108.

10. Rogers, T. H. and W. K. Chinn; *Uses and Abuses of Aluminum in Wooden-Hulled Aluminum Frame Minesweeper;* Corrosion, Vol. XV, No. 8, Aug. 1959; Houston. pp. 21-26.

11. Rohan, Jack; *Yankee Arms Maker (The Incredible Career of Samuel Colt);* Harpers & Bros., 1935; New York.

12. Roland, Alex Frederick; *A Triumph of Natural Magic: The Development of Underwater Warfare in the Age of Sail, 1571-1865;* Dissertation, Duke University, Ph.D., 1974.

13. Rollman, G.; *The War in the Baltic;* Military Publishing House, Vol. II, (1915) 1935; Moscow.

14. von Romocki, S. J.; *Geschichte der Sprengstoffchemie, der Sprengtechnik und des Torpedowesens;* Janecke Bros., Part I, Chap. XI, 1895; Hanover. pp. 323-337.

15. Roosevelt, Theodore; *The Naval War of 1812;* G. P. Putnam's Sons, 3rd Edition, 2 volumes, 1900; New York and London.

Roscoe, Theodore; *This is Your Navy*; U.S. Naval Institute, 1950; Annapolis.   1.
    pg. 539.

Roscoe, Theodore; *United States Destroyer Operations in World War II*; U.S.   2.
    Naval Institute, 1953; Annapolis, Maryland.

Roscoe, Theodore; *United States Submarine Operations in World War II*; U.S.   3.
    Naval Institute, 1949; Annapolis, Maryland.

Roskill, Stephen Wentworth; *Naval Policy Between the Wars. The Period of*   4.
    Anglo-American Antagonism 1919-1929, Part 1.

Roskill, Stephen Wentworth; *The Strategy of Sea Power*; Collins, 1962; London.   5.

Roskill, Stephen Wentworth; *The War at Sea 1939-1945 (History of the Second*   6.
    World War); Her Majesty's Stationery Office, 3 volumes, 1954; London.

Rothbotham, W. B.; *Robert Fulton's Turtle Boat*; U.S. Naval Institute Pro-   7.
    ceedings, Vol. LXII, No. 12, Dec. 1936; Annapolis, Maryland. pp. 1746-
    1749.

Rougerson, C.; *Les Mines Magnetiques*; L'Illustration, Vol. CCIV, No. 5050,   8.
    Dec. 16, 1939; Paris. pp. 434-436.

Rough Passage to a Date with Destiny; Sentinel, Vol. III (Canadian Forces   9.
    Sentinel), May 1967; Ottawa, Canada. pg. 46.

Routledge, Robert; *Discoveries and Inventions of the Nineteenth Century*   10.
    12th Edition; George Routledge, 1898; London. pp. 167-181.

Rowe, O.; *Garrison Artillery Warfare*; Journal of the U.S. Artillery, Vol.   11.
    VI, No. 1, Jul-Aug. 1896; Ft. Monroe, Virginia. pp. 68-80.

Rowland, Buford and William B. Boyd; *U.S. Navy Bureau of Ordnance in World*   12.
    War II; Bureau of Ordnance, Department of the Navy, U.S. Government
    Printing Office, 1953; Washington, D.C.

Royal Canadian Navy Gets 5 Bay Class Sweepers; Air Force Times, Vol. XVII,   13.
    Jul. 6, 1957; Washington, D.C. pg. E-6.

(Royal George); A Narrative of the Loss of the Royal George at Spithead,   14.
    August 1781; including Tracey's Attempt to Raise Her in 1783. Also
    Col. Pasley's Operations in Removing the Ship by Gunpowder in 1839-
    40-41. Bound in the Wood of the Wreck. Fifth Edition, S. Horsey,
    Senior, 1842; Portsea.

1. Ruge, Friedrich O.; *Comments and Discussion on "Russian Mines on the Danube"*; U.S. Naval Institute Proceedings, Vol. XCIII, No. 4, Apr. 1967; Annapolis, Maryland. pg. 117.

2. Ruge, Friedrich O.; *German Mine Sweepers in World War II*; U.S. Naval Institute Proceedings, Vol. LXXVIII, No. 9, Sept. 1952; Annapolis, Maryland. pp. 994-1003.

3. Ruge, Friedrich O.; *German Naval Strategy Across Two Wars*; U.S. Naval Institute Proceedings, Vol. LXXXI, No. 2, Feb. 1955; Annapolis, Maryland. pp. 153-166.

4. Ruge, Friedrich O.; *Minen an der Amerikaneschen Kuste*; Marine Rundschau, Vol. XXXIX, 1934; Berlin. pp. 392-400.

5. Ruge, Friedrich O.; *Der Seekrieg, The German Navy's Story 1939-1945*; U.S. Naval Institute, 1957; Annapolis, Maryland.

6. Ruge, Friedrich O.; *Torpedo und Minekrieg*; T. F. Lehmans, 1940; Berlin. pp. 30-67.

7. Ruge, Friedrich O.; *Die Verwendung der Mine im Seekriege 1914 bis 1918 (Ihre Erfolge und Mizerfolge)*; Marine Rundschau, Vol. XXXII, No. 6, Mittler und Sohn, Jun. 1927; Berlin. pp. 257-266.

8. Ruge, Friedrich O.; *With Rommel Before Normandy*; U.S. Naval Institute Proceedings, Vol. LXXX, No. 6, Jun. 1954; Annapolis, Maryland. pp. 612-619.

9. Rushmore, D. B.; *General Electric Review, 1917*; Bibliography of the Literature of Submarines, Mines and Torpedoes, Vol. XX. pp. 675-685.

10. *Russia*; Army and Navy Journal, Vol. XIX, No. 17, Nov. 26, 1881; New York. pg. 369.

11. *Russian Prizes for Torpedo Defense*; Engineering, Vol. XLI, Feb. 26, 1886; London. pg. 210.

12. *Russian Torpedoes*; Engineering, Vol. XXIV, Jul. 6, 1877; London. pg. 19.

13. *Russiske Helvedes-Maskiner i Østersøen i 1855*; Tidsskrift for Sovaesen, Vol. XXVII, 1855; Copenhagen. pg. 417.

14. *Russo-Japanese War 1903-1905*; Office of the Chief of Staff, War Department Document No. 279.

Ryan, Cornelius; *The Longest Day*; Simon & Schuster, Inc., 1959;   1.
    New York.

Ryan, L. S.; *Mine Commander's Board*; Journal of the U.S. Artillery, Vol.   2.
    XXXIII, No. 1, Jan-Feb. 1910; Ft. Monroe, Virginia. pp. 33-37.

Saar, Charles W.; *Offensive Mining as a Soviet Strategy*; U.S. Naval In-   3.
    stitute Proceedings, Vol. XC, No. 8, Aug. 1964; Annapolis,
    Maryland. pp. 42-51.

Samarov, A. A. and F. A. Petrov; *Development of Mine Materiel in the*   4.
    Russian Navy; Naval Publishing House, 1951; Moscow.

Sanders, Harry; *The First American Submarine*; U.S. Naval Institute Pro-   5.
    ceedings, Vol. LXII, No. 12, Dec. 1936; Annapolis, Maryland.
    pp. 1743-1745.

Sargent, Nathan; *Admiral Dewey and the Manila Campaign*; Naval History   6.
    Foundation, 1947.

de Sarrepont, Major H. (Pseud. de Hennebert, Lt. Col. Eugene); *Les*   7.
    Torpilles; Imprimerie en Libraire Militaire, J. Dumaine, 1874; Paris.
    xii, 320 p. et figures. [Extrait du "Journal des sciences militaires".]

de Sarrepont, Major H. (Pseud. de Hennebert, Lt. Col. Eugene); *Les*   8.
    Torpilles; Baudoin, 1883; Paris.

de Sarrepont, Major H. (Pseud. de Hennebert, Lt. Col. Eugene); *Les*   9.
    Torpilles; Dumaine, 1880; Paris. vii, 574 p.

Saunders, M. G. (Editor); *The Soviet Navy*; Praeger, 1958; New York.   10.

Saunders, M. G. (Editor); *Soviet Navy*; Weidenfeld and Nicolson, 1958;   11.
    London.

Sauvaire-Jourdan, F.; *Les Mines Sous-Marines et leur Emploi par les Sous-*   12.
    Marines; La Nature, Vol. LV, Part 2, Dec. 15, 1927; Paris.
    pp. 549-552.

Savage, Carlton; *Policy of the United States Toward Maritime Commerce in*   13.
    War; U.S. Government Printing Office, Vol. II, 1914-1918, 1936;
    Washington, D.C.

Scharf, J. Thomas; *History of the Confederate States Navy from its Organi-*   14.
    zation to the Surrender of its Last Vessels; Rodgers & Sherwood, 1887;
    New York.

Scheer, Reinhard von; *Germany's High Sea Fleet in the World War*; Cassell   15.
    and Co., Ltd., 1920; London, New York and Toronto.

1. Scheer, Reinhard von; *Germany's High Sea Fleet in the World War*; Peter Smith, 1934; New York.

2. Scheidnagel, D. Leopold; *Extracts from a Treatise upon Defensive Submarine Mining*; U.S. Engineers Professional Papers, Paper No. 2 (translated by 1st Lt. Fred Abbot), Press of the Battalion of Engineers, 1881-82; Willets Point, New York.

3. Scheidnagel, D. Leopold; *Minas Hidraulicas Defensivas*; Libreria de Francisco Iravedra, 1866; Madrid.

4. (Scheliha), *One von Scheliha*; Army and Navy Journal, Vol. VI, No. 38, May 8, 1869; New York. pg. 600.

5. von Scheliha, Viktor Ernst Karl Rudolf, *A Treatise on Coast-Defence*; E. and F. N. Spon, 1868; London.

6. Schell, F. H.; *Submarine Warfare During the Civil War and Its Modern Development*; Leslie's Weekly, Vol. LXXXVI, Apr. 28, 1898; New York. pp. 264-266.

7. Schofield, B. B.; *The Royal Navy Today*; Oxford University Press, 1960; London.

8. *School of Submarine Defense*; Army and Navy Journal, Vol. XLII No. 52, Aug. 26, 1905; New York. pg. 1420.

9. *The Schoole - HMS Vernon*; Navy, Vol. LXXIV, No. 6, Jun. 1969; London. pp. 202-204.

10. Schriebershofen, Max; *Seeminen ...*; Stuck, 1915; Leipzig. 29 p.

11. Schull, Joseph; *The Far Distant Ships*; Minister of National Defense, 1952; Ottawa.

12. Schultz, J. H.; *Flaadens Virksomhed under Verdens Kriegen 1914-1919*; 1920; Copenhagen.

13. Schultz, M.; *They Hunt for Floating Death; Viet Congs Explosive Mines*; Popular Mechanics, Vol. CXXIX, No. 4, Apr. 1968; Chicago. pp. 86-89.

14. Schurman, D. M.; *Education of a Navy. Development of British Naval Strategic Thought 1867-1914*; University of Chicago Press, 1965; London.

15. Schwarte, M.; *Die Technik in Weltkriege*; Mittler und Sohn, 1920; Berlin.

Scott, James B. (Editor); *Instructions to the American Delegates to the Hague Peace Conference and their Official Report;* Oxford University Press, 1916; New York. pg. 111. 1.

Scott, James B. (Editor); *The Reports of Hague Peace Conferences of 1899 and 1907;* Clarendon Press, 1917; Oxford. 2.

Scoville, Herbert, Jr.; *"Beyond SALT One";* Foreign Affairs, Vol. L, No. 3, Apr. 1972. 3.

Scoville, Herbert, Jr.; *"Missile Submarines and National Security";* Scientific American, Vol. CCVI, No. 6, Jun. 1972. 4.

Sea Eggs: Processes in the Making of Submarine Mines; Illustrated London News, Vol. CIC, Dec. 6, 1941; London. pg. 731. 5.

A Sea Fight in the Adriatic, Destroying a Minefield Under the Guns of the Enemy; Scientific American, Vol. CXVII, No. 2, Jul. 14, 1917; New York. pp. 24-25, 34. 6.

Seagoing Soldiers; Newsweek, Vol. XIX, No. 2, May 18, 1942; Washington, D.C. pg. 32. 7.

Sears, James H.; *The Coast in Warfare, Part I;* U.S. Naval Institute Proceedings, Vol. XXVII, No. 3, Sept. 1901; Annapolis, Maryland. pp. 449-527. 8.

Sears, James H.; *The Coast in Warfare, Part II;* U.S. Naval Institute Proceedings, Vol. XXVII, No. 4, Dec. 1901; Annapolis, Maryland. pp. 649-712. 9.

Sea: Third Weapon; Newsweek, Vol. XV, No. 7, Feb. 12, 1940; Washington, D.C. pp. 28-29. 10.

The Sea Torpedo; Army and Navy Journal, Vol. VI, No. 34, Apr. 10, 1869; New York. pg. 534. 11.

Seawright, Murland W.; *Prepare to Sweep Mines;* U.S. Naval Institute Proceedings, Vol. XCVI, No. 1, Jan. 1970; Annapolis, Maryland. pp. 54-59. 12.

Seeminen; Ueberall, Vol. I, 1901; Berlin. pp. 393, 408. 13.

Seeminen; Ueberall, Vol. XIV, No. 5 and 7, 1912; Berlin. pp. 322-327 and 475-480. 14.

Seeminen; Ueberall, Vol. XVII, No. 9, 1915; Berlin. pp. 546-549. 15.

1. *Seeminen und Mineleger in der Frankosischen Kriegsmarine;* Mitterlungen aus dem Gebiete des Seewesens, Vol. XL, 1912; Pola. pp. 958-965.

2. Semon, H. W.; *History of Switch Horn Development at NOL;* Naval Ordnance Laboratory, Apr. 18, 1946; White Oak, Maryland.

3. Sendall, W. R.; *Exploits of the Limpeteers;* The Navy, Vol. LII, No. 10, Oct. 1947; London. pg. 384.

4. *Setting Traps for Enemy Ships;* Popular Mechanics, Vol. LXXIII, No. 5, May 1940; Chicago. pp. 706-709.

5. Sessions, George Perry; *They Fight the Axis Devil Fish;* Saturday Evening Post, Vol. CCVII, No. 22, Nov. 25, 1944; Philadelphia. pp. 11, 100-101.

6. Settle, Stuart Williston; *Confederate Use of Mine Warfare in Charleston Harbour, 1861-1865;* Naval Schools, Mine Warfare, Staff Officers Course Term Paper, May 7, 1966; Charleston, South Carolina.

7. *Seventh Fleet Mine Warfare Officer Discusses Operating Against North Viet-Nam;* Navy, Vol. X, No. 5, May 1967; Washington, D.C. pg. 12.

8. Seves, Lt. de vais.; *Cours de Torpilles;* Défense fixe, Imprimerie Nationale, 1900; Paris. 90 p. [Ministère de la Marine.]

9. Sewell, John Stephen; *Electricity in its Application to Submarine Mines;* Transactions of the American Institution of Electrical Engineers, Oct. 1902; New York.

10. Sewell, John Stephen; *Electricity in its Application to Submarine Mines;* U.S. Naval Institute Proceedings, Vol. XXVIII, No. 3, Sept. 1902; Annapolis, Maryland. pp. 708-711.

11. *Sewing Up the Subs (Abridged from Adm. Sims, Victory at Sea);* All Hands, No. 516, Jan. 1960; Washington, D.C. pp. 59-63.

12. *Shallow Water Sweepers;* All Hands, No. 343, Apr. 1953; Washington, D.C. pg. 36.

13. Shaw, Frances B.; *A Collection of Papers on Mining;* U.S. Naval Ordnance Laboratory, Jan. 3, 1955; White Oak, Maryland.

14. Shearing, D.; *Diesel AM Class Minesweeper;* Diesel Progress, Vol. XI, No. 7, Jul. 1945; New York. pp. 64-65.

Sheppard, William; *Dismal Spit and Her Mackerel Taxis;* U.S. Naval Institute   1.
    Proceedings, Vol. LXX, No. 10, Oct. 1944; Annapolis, Maryland. pp.
    1253-1257.

Sheridan, M.; *Planting Mines by the Army;* Travel, Vol. LXXX, No. 2, Feb.   2.
    1943; Floral Park, New York. pp. 18-21 and 33-34.

Sherwood, John; *Divers Win Judgement Smithsonian is Torpedoed;* Washington   3.
    Evening Star, Vol. CXX, No. 364, Friday, Dec. 29, 1972; Washington, D.C.
    pg. B-1.

Ships Under-Water Eye Sights Hidden Mines; Popular Mechanics Magazine,   4.
    Vol. XXVI, 1916; Chicago. pg. 161.

Showdown Over Vietnam; U.S. News and World Report, Vol. LXXII, No. 21,   5.
    May 22, 1972; Washington, D.C. pp. 16-18.

Shulman, O. V. and B. A. Shimanyak; *The Mine as a Weapon Under Contemporary*   6.
    Conditions; Morskoy Sbornik, No. 12, 1967; Moscow. pp. 39-43.

"Sic Fidem Teneo"; *The Future of Submarine Mining;* United Service Magazine,   7.
    Vol. XXXI, No. 920, Jul. 1905; London. pp. 361-364.

Sigsbee, Charles Dwight; *Personal Narrative of the Maine;* Century Magazine,   8.
    Vol. LVII, No. 3, Jun. 1899; New York. pp. 373-394.

Sigtevenklens Bestemmelse m.v.; Tidsskrift for Sovaesen, Vol. CIII, 1932;   9.
    Copenhagen. pg. 53.

Silas, Ellis; *Infancy of the Mine H.M.S. Merlin in the Crimean War;* Repro-   10.
    duction of a painting, The Navy, Vol. LVI, No. 12, Dec. 1951; London.
    pg. 308.

Simon, Leslie Earl; *German Research in World War II;* J. Wiley and Sons,   11.
    Chapman and Hall, Ltd., 1947; New York, London.

Sims, William Sowden; *American Mine Barrage in the North Sea;* Worlds Work,   12.
    Vol. XL, Jun. 1920; New York. pg. 72.

Sims, William Sowden; *Victory at Sea;* Doubleday, Page & Co., 1920; Garden   13.
    City, New York.

Siney, Marion C.; *The Allied Blockade of Germany 1914-1916;* University of   14.
    Michigan Press, 1957.

(Sinker of the "Bouvet", "Irresistible" and "Ocean"); Illustrated London   15.
    News, Vol. CXLVI, 1915; London. pg. 461.

1. Skerrett, R. G.; *Acoustic Mines Fired by Pneumatic-Tool Beats;* Compressed Air Magazine, Vol. LI, Jun. 1946; Philipsburg, New Jersey. pp. 160-161.

2. *Sketch of a Torpedo Picked Up in Marganza Bend, Mississippi River by the Crew of the USS Lafayette;* Frank Leslie's Illustrated Newspapers, Vol. XX, 1865; New York. pg. 189.

3. *Slaebeminer;* Tidsskrift for Sovaesen, Vol. C, 1929; Copenhagen. pg. 38.

4. Sleeman, Charles William; *Torpedoes and Torpedo Warfare;* Griffin & Co., 1880; Portsmouth.

5. Sleeman, Charles William; *Torpedoes and Torpedo Warfare;* Second Edition, Griffin & Co., 1889; Portsmouth. 354 p.

6. Smith, A. W.; *Greenport Basin and Construction Company Turning Out Mine-sweepers;* Rudder, Vol. LVIII, Jan. 1942; New York. pp. 60-61.

7. Smith, C. J.; *An Outline of Russian History;* Office of the Chief of Naval Operations, 1958; Washington, D.C.

8. Smith, F. E. and N. W. Sibley; *International Law as Interpreted During Russo-Japanese War;* Boston Books, 1905; Boston.

9. Smith, J. Bucknall; *Wire, Its Manufacture and Uses;* John Wiley & Sons, 1891; New York.

10. Smith, S. E.; *The United States Navy in World War II;* Morrow & Co., Inc., 1966; New York.

11. Smythe, Augustine T.; *Year Book, City of Charleston, 1907;* Walker, Evans and Cogswell Company, 1908; Charleston, South Carolina.

12. Snow, Chester R. and Harold G. Douglas; *Mine Efficiency From Store House to Dock;* Journal of the U.S. Artillery, Vol. XLVI, No. 2, Sept-Oct. 1916; Ft. Monroe, Virginia. pp. 319-332.

13. Sokol, Anthony Eugene; *The Imperial and Royal Austro-Hungarian Navy;* U.S. Naval Institute, 1968; Annapolis, Maryland.

14. Sokol, Anthony Eugene; *Torpedo Boat Carriers;* U.S. Naval Institute Proceedings, Vol. LXVIII, No. 11, Nov. 1942; Annapolis, Maryland. pp. 1586-1590.

15. Sokol, Hans H.; *La Marine Austro-Hongroise dans la Guerre Mondiale, 1914-1918;* 1933; Paris.

Sokol, Hans H.; *Österreich-Ungarns Seekrieg, 1914-1918*; 1933; Vienna. 1.

The Solent Mine Field; Fairplay, Vol. XLVII, Oct. 25, 1906; London. pg. 616. 2.

Soley, James Russell; *The Navy in the Civil War: The Blockade and the* 3.
    Cruisers; Sampson Low, 1898; London.

Some Aspects of Mine Warfare; International Defense Review, Interavia 4.
    Home Office, 1/1969; Geneva.

Sominervaesenet 1878-1 April 1953. Tilboglblik Over Perioden 1928-1953; 5.
    Tidsskrift for Sovaesen, Vol. CXXIV, 1953; Copenhagen. pg. 433.

Soulage, C. C.; *The Defense of Ports and Harbors with Electrical Torpedoes;* 6.
    Scientific American Supplement, Vol. XIX, No. 480, Mar. 14, 1885;
    New York. pp. 7655-7656.

Southall, Ivan; *Dix-sept Secondes pour Survivre*; Colman-Levy, 1963; 7.
    (Deminage) Paris. 314 p.

Southall, Ivan; *Softly Tread the Brave*; Angus and Robertson, 1960; London. 8.

The Soviet Russians as Opponents at Sea; ONI Translation No. 287 a, b, 9.
    c, d (4 volumes). Analysis of German and Russian Naval Operations
    in the Second World War.

Sowing and Destroying Mines; U.S. Naval Institute Proceedings, Vol. XXXV, 10.
    No. 2, Jun. 1909; Annapolis, Maryland. pg. 596.

Sowing and Reaping Mines; United Service Gazette, Vol. CLXIII, Nov. 5, 1914; 11.
    London. pp. 333-334.

Spaerringen af de andske Geennemsylmys farvande i August 1914; Tidsskrift 12.
    for Sovaesen, Vol. CI, Nov-Dec. 1930; Copenhagen. pp. 565-662.

Sparks, Jared; *The Library of American Biography;* Little & Brown, Vol. XI, 13.
    Second Series, 1846; Boston.

Speeding Up Sea Defenses; Army and Navy Journal, Vol. LIV, No. 39, 14.
    May 26, 1917; New York. pg. 1257.

Spindler, Arno; *Der Handelskrieg mit U-Booten;* Dritter Band (Volume III) 15.
    of *Der Krieg Zur See 1914-18;* Oktober 1915 bis January 1917, Mittler
    & Sohn, 1934; Berlin.

1. (Spindler); Journal of Royal United Service Institution, Vol. LXXIX, No.
    515, Aug. 1934; London. pg. 646. [Book review.]

2. Spindler, Arno; *Value of the Submarine in Naval Warfare*; U.S. Naval Institute
    Proceedings, Vol. LII, No. 5, May 1926; Annapolis, Maryland. pp. 835-
    854.

3. Sprout, Harold and Margaret; *Toward A New Order of Sea Power*; University of
    Princeton Press, 1940; Princeton, New Jersey.

4. *Staaltraadsbarrikader*; Tidsskrift for Sovaesen, Vol. LXV, 1894; Copenhagen.
    pg. 264.

5. Stadtlander, Gerd; *Die Verwendung von Minen im Seekrieg*; G. H. Nolte, 1938;
    Dusseldorf.

6. Stafford, Edward P.; *The Far and the Deep*; Arthur Baker, Ltd., 1968; London.

7. Stebbins, John; *The Mine Sweeping/Fishing Vessel*; U.S. Naval Institute Pro-
    ceedings, Vol. XLVII, No. 3, Mar. 1971; Annapolis, Maryland. pp. 91-92.

8. von Stengel, K.; *Die Entwicklung des Kriegsrechts im Allgemeinen und das
    Seekriegsrecht im besonderen*; Marine Rundschau, 1905; Berlin.
    pp. 275-294, 416-440.

9. Sterling, Y.; *Fighting the Submarine Mine*; Popular Science, Vol. CXXIX,
    No. 4, Oct. 1941; New York. pp. 102-108.

10. Stern, Philip von Doren; *The Confederate Navy a Pictorial History*; Doubleday
    and Company, 1962; Garden City, New York.

11. Steward, Harding; *Notes on Submarine Mines, Commonly Called Torpedoes*;
    Jackson and Son, 1866; Woolwich.

12. Stockton, Charles Herbert; *The Use of Submarine Mines and Torpedoes in
    Time of War*; The American Journal of International Law, Vol. II,
    No. 2, Apr. 1908; New York. pp. 276-284.

13. Stokes, Anson Phelps; *Memorials of Eminent Yale Man*; Yale University Press,
    Vol. II, 1914; New Haven.

14. Stokes, Donald; *Les Mines Magnétiques Auraient pu Faire Perdue la Guerre
    aux Allies*; "Historia" Aurel, 1954; Paris. 6 p.

15. Stone, Julius; *Legal Controls of International Conflict*; Rinehart, 1954;
    New York. pg. 584.

(Stotherd); Army and Navy Journal, Vol. XI, No. 11, Foreign Items, Oct. 25, 1873; New York. pg. 74. 1.

(Stotherd); Army and Navy Journal, Vol. XI, No. 20, Dec. 27, 1873; New York. pg. 313. 2.

Stotherd, Richard Hugh; *Defensive Submarine Warfare*; Journal of the Royal United Service Institution, Vol. XV, No. 65, 1872; London. pp. 705-733. 3.

Stotherd, Richard Hugh; *Notes on Defence by Submarine Mines*; 2nd Edition, Revised and Enlarged, 1873; Brompton, Kent. 4.

Stotherd, Richard Hugh; *Report on the German Torpedo Establishment at Kiel and Wilhelmshafen*; Nov. 25, 1874; Whitehall. 5.

Stotherd, Richard Hugh; *Torpedoes, Offensive and Defensive*; U.S. Government Printing Office, 1872; Washington, D.C. [Binding may be titled Submarine Mines.] 6.

Strange Mine Detonating Egg - Crates for Scrap After War Service; Illustrated London News, Vol. CCX, No. 5622, 1947; London. 7.

"Strategicus"; *War Surveyed: The War at Sea*; Spectator, Vol. CLXIII, Dec. 1, 1939; London. pg. 767. 8.

Strauss, J.; *Pulling the Teeth of War*; Forum, Vol. XLIII, Jan. 1920; New York. pp. 57-68. 9.

Stroh, H.; *Mines et Torpilles*; Armand Colin, 1924; Paris. 10.

Stryker, Lyal M.; *Comments on "The Mine as a Tool of Limited War"*; U.S. Naval Institute Proceedings, Vol. LXIII, No. 7, Jul. 1967; Annapolis, Maryland. pg. 107. 11.

Stryker, Lyal M.; *Corrections to "Naval Mine Warfare"*; U.S. Naval Institute Proceedings, Vol. LXXXVI, No. 7, Jul. 1960; Annapolis, Maryland. pg. 100. 12.

Sturdee, Sir Fredrick C. D.; *The Changes in the Conditions of Naval Warfare, etc.*; Journal of the Royal United Service Institution, Vol. XXX, No. 134, 1886; London. pp. 367-418. 13.

The Submarine and Kindred Problems; The Engineer, Vol. CXXIV, Oct. 19, 1917; London. pp. 329-330. 14.

Submarine Harbor Defense Mine in Baltimore Harbor; Engineering News, Vol. XLI, No. 8, Feb. 23, 1899; New York. pg. 114. 15.

1.   *Submarine Infernal Machines;* Scientific American, Vol. V, No. 7, Aug. 17,
        1861; New York. pg. 101.

2.   *The Submarine Mine;* Scientific American, Vol. XC, No. 17, Apr. 23, 1904;
        New York. pg. 330.

3.   *The Submarine Mine;* The Times Literary Supplement, No. 2475, Jul. 8, 1949;
        London. pg. 442.

4.   *Submarine Mine and Home Coast Defence;* Journal of the Military Service In-
        stitution, Vol. XXIX, No. 113, Sept. 1901; Governor's Island, New York.
        pg. 282.

5.   *Submarine Mine Defense of Coast Fortresses;* Journal of the U.S. Artillery,
        Vol. XXXVII, No. 2, Mar-Apr. 1912; Ft. Monroe, Virginia. pp. 196-208.

6.   *Submarine Mine Defense of Coast Fortresses;* Journal of the U.S. Artillery,
        Vol. XXXVII, No. 3, May-Jun. 1912; Ft. Monroe, Virginia. pg. 347.

7.   *Submarine Mine Defense of Coast Fortresses;* Journal of the U.S. Artillery,
        Vol. XXXVIII, No. 1, Jul-Aug. 1912; Ft. Monroe, Virginia. pp. 59-70.

8.   *The Submarine Mine in Modern Naval Warfare;* The Navy, Vol. XXXV, No. 12,
        Dec. 12, 1930; London. pg. 370.

9.   *Submarine Mine-Layer;* Literary Digest, Vol. LIII, No. 18, Oct. 28, 1916;
        New York. pp. 1104-1105.

10.   *Submarine Mine Layers;* Scientific American Supplement, Vol. LXVII, No. 1739,
        May 1, 1909; New York. pg. 278.

11.   *Submarine Minelayers;* U.S. Naval Institute Proceedings, Vol. XLI, No. 6,
        Nov-Dec. 1915; Annapolis, Maryland. pp. 2019-2020.

12.   *Submarine Mines;* Army and Navy Journal, Vol. LII, No. 19, Jan. 9, 1915;
        New York. pg. 584.

13.   *Submarine Mines;* Corps of Engineers, U.S.A.; Professional Papers, No. 23,
        Battalion Press, Willets Point, New York.

14.   *Submarine Mines;* Scientific American Supplement, Vol. LII, No. 1334, Jul. 27,
        1901; New York. pp. 21386-21387.

15.   *Submarine Mines and Mining;* Army and Navy Journal, Vol. XLV, No. 44, Jul. 4,
        1908; New York. pg. 1222.

Submarine Mines and Their Place in Naval Warfare; Scientific Australian,   1.
    Vol. XX, No. 2, Dec. 1914; Melbourne. pp. 40-41.

Submarine Mines for Naval Work; Journal of the U.S. Artillery, Vol. XXXVII,   2.
    No. 1, Jan-Feb. 1912; Ft. Monroe, Virginia. pp. 107-108.

Submarine Mines on the High Seas; Scientific American, Vol. XC, No. 23,   3.
    Jun. 4, 1904; New York. pg. 434.

Submarine Mining; Engineering, Vol. XXIII, Mar. 4, 1877; London. pg. 213.   4.

Submarine Mining and Torpedo Warfare; Scientific American, Vol. LXXVIII,   5.
    No. 10, Mar. 5, 1898; New York. pp. 149-150.

Submarine Mining, Rough Notes of Seven Lectures on; RE School of Military   6.
    Engineering, 1877; Chatham.

Submarine Torpedoes - Infernal Machines; Scientific American, Vol. VI,   7.
    No. 11, Mar. 15, 1862; New York. pp. 164-165.

Submarine War; Engineering, Vol. IX, Feb. 18, 1870; London. pp. 104-105.   8.

Submarine Warfare; Army and Navy Journal, Vol. VII, No. 9, Oct. 16, 1869;   9.
    New York. Front page and pg. 125.

Submarine Warfare; Army and Navy Journal, Vol. VII, No. 16, Dec. 4, 1869;   10.
    New York. pp. 237-238.

Submarine Warfare; Scientific American, Vol. X, No. 18, Apr. 30, 1864;   11.
    New York. pg. 282.

Submarine Warfare; Van Nostrand's Engineering Magazine, Vol. II, No. 16,   12.
    Apr. 1870; New York. pp. 409-412.

Submarines and Mine-Fields; United Service Gazette, Vol. CLXIII, Dec. 24,   13.
    1914; London.

Submarines and Mine Sweepers; Army and Navy Journal, Vol. LVI, No. 17,   14.
    Dec. 28, 1918; New York. pg. 609.

Submarines and Mines; U.S. Naval Institute Proceedings, Vol. XXXVI, No. 4,   15.
    Dec. 1910; Annapolis, Maryland. pp. 1192-1194.

Suddath, Thomas H.; *The Role of the United States Navy in Mine Warfare;*   16.
    U.S. Naval Institute Proceedings, Vol. XCI, No. 9, Sept. 1965;
    Annapolis, Maryland. pg. 108.

1. Sueter, M. F.; *The Evolution of the Submarine Boat, Mine and Torpedo;* J. Griffin & Co., 1907; Portsmouth.

2. Sullivan, Barry W.; *Mine Countermeasures in the U.S. Civil War;* Naval Schools, Mine Warfare, Staff Officers Term Paper, May 26, 1969; Charleston, South Carolina.

3. Sunderland, Archibald H.; *The S.S. "Noordam" and a Mine;* Journal of the U.S. Artillery, Vol. XLV, No. 3, May-Jun. 1916; Ft. Monroe, Virginia. pp. 314-315.

4. Sundley, Sir Robert; *Mine Laying by Bomber Command;* Air Power, Vol. VII, Summer 1960; London. pp. 261-263.

5. *Sunk by a One in a Million Chance; The Belgian Steamer Prinses Astrid;* Illustrated London News, Vol. CCXV, Jul. 2, 1949; London. pp. 1-3.

6. *Survey of Cargo Reinsurance Exchange Shows Mine Hazard Still Serious;* Weekly Underwriter, Vol. CLXXV, Jul. 7, 1956; New York. pg. 17.

7. (Swedish Defence Reform 1914); United Service Magazine, New Series, Vol. L, No. 1033, Dec. 1914; London. pp. 287-289.

8. *Sweeping Drill;* All Hands, No. 531, Apr. 1961; Washington, D.C. pg. 9.

9. *Sweeping for Mines;* Illustrated London News, Vol. CCXXV, Oct. 30, 1954; London. pg. 738.

10. *Sweeping Sudden Death;* Popular Mechanics, Vol. LXXXV, No. 1, Jan. 1946; Chicago. pp. 28-34.

11. *Sweeping the Seas Clear of Mines;* Worlds Work, Vol. XXXIV, Nov. 1919; London. pp. 508-514.

12. *The Sweeps Sweep On;* All Hands, No. 342, Sept. 1945; Washington, D.C. pp. 22-25.

13. *.wimming Mine;* Engineering Magazine, Vol. L, Oct. 1915; London. pg. 122.

14. *Swimming Mine;* Journal of the U.S. Artillery, Vol. XLIV, No. 3, Jul-Dec. 1915; Ft. Monroe, Virginia. pp. 382-383.

15. *A Symposium ...;* Nature, Vol. CLVII, No. 3987, Mar. 30, 1946; London. pg. 404. [On degaussing.]

Symposium of Papers on Degaussing; Electrical Review, Vol. CXXXVIII, No. 3568, Apr. 12, 1946; London. pp. 565-566.   1.

"Taffrail"; *Endless Story*; Hodder & Stoughton, Ltd., 1932; London.   2.

"Taffrail"; *The Minewatchers*; The Navy, Vol. LX, No. 2, Feb. 1955; London. pp. 29-30.   3.

"Taffrail"; *Salute to the Sweepers*; The Navy, Vol. LII, No. 2, Feb. 1947; London. pp. 51-53.   4.

"Taffrail"; *Sowing Death at Sea*; Saturday Evening Post, Vol. CCXII, No. 44, Apr. 27, 1940; Philadelphia. pg. 29.   5.

"Taffrail" on the Sea War Action; The Times Literary Supplement, No. 2013, Aug. 31, 1940; London. pg. 419.   6.

Taking Up the Mines (Excerpts from the "Northern Barrage"); All Hands, No. 471, May 1956; Washington, D.C. pp. 59-63.   7.

Talks in Haiphong on Mine Clearing Will Open Today; Washington Post, Vol. XCVI, No. 62, Monday, Feb. 5, 1973; Washington, D.C. Front page.   8.

Tamura, Kyuzo; *History of Japanese Minesweeping, 1937-1947*; Minesweeping Division, Second Demobilization Bureau, Jun. 1948; Sasebo.   9.

(Target Report) Ordnance Target: Countermeasures and Defensive Organization of Japanese Against U.S. Mines; U.S. Naval Technical Mission to Japan, O-03, Jan. 3, 1946.   10.

(Target Report) Ordnance Target: Japanese Mines; U.S. Naval Technical Mission to Japan, O-04, Jan. 11, 1946.   11.

(Target Report) Ordnance Target: Japanese Minesweeping Gear and Equipment; U.S. Naval Technical Mission to Japan, S-28, Jan. 1946.   12.

(Target Report) Ordnance Target: Japanese Naval Mining Organization and Operational Techniques; U.S. Naval Technical Mission to Japan, O-05, Jan. 2, 1946.   13.

(Target Report) Ship and Related Targets: Japanese Degaussing; U.S. Naval Technical Mission to Japan, S-37, Dec. 25, 1945.   14.

Taylor, Edmond; *The Fall of the Dynasties: The Collapse of the Old Order 1905-1922*; Doubleday, 1963.   15.

1. Taylor, William D., Captain, USN (Ret.); *Some Thoughts on the Employment of Countermeasures in "Peacetime"*; Working Paper X1116, Study on Countermeasures in Fleet Defense, Aug. 10, 1973.

2. Telberg, V.; *Russian-English Dictionary of Nautical Terms*; Telberg Book Corp., New York.

3. *The Telephone and the Torpedo*; Journal of the Royal Society of Arts, Vol. XXVI, 1878; London. pp. 887-888.

4. *Telephonic Torpedoes*; Army and Navy Journal, Vol. XXI, No. 6, Sept. 14, 1878; New York. pg. 87.

5. Thach, John W.; *The ASW Navy of the Seventies*; U.S. Naval Institute Proceedings, Vol. LXXXIX, No. 1, Jan. 1963; Annapolis, Maryland. pg. 62.

6. Thacher, James; *The American Revolution*; American Subscription Publishing, 1860; New York.

7. Thacher, James; *A Military Journal During the American Revolutionary War From 1775 to 1783*; Cottons and Barnard, 2nd Edition, Revised and Corrected, 1827.

8. *The Theory and Practice of Degaussing*; Engineering, Vol. CLXI, Apr. 12, 1946; London. pg. 343.

9. Thomas, Hugh; *The Spanish Civil War*; Eyre and Spottiswoode, 1961; London.

10. Thomas, Lowell; *Raiders of the Deep*; Garden City Publishing Co., 1940.

11. Thomazi, Auguste; *La Marine Française 1914-1918*; Payot, 1919, 1925, 1926, 1933; Paris.

12. Thompson, R. W.; *At Whatever Cost*; Coward-McCann, Inc., 1957; New York.

13. Thomson, David W.; *David Bushnell and the First American Submarine*; U.S. Naval Institute Proceedings, Vol. LXVIII, No. 2, Feb. 1942; Annapolis, Maryland. pp. 176-186.

14. Thurston, Robert H.; *Robert Fulton (His Life and its Result)*; Dodd, Mead & Company, 1891; New York.

15. *The Times Documentary History of the War*; London. The Times Publishing Co., Ltd., Vol. XI, Naval Part 4, 1920. pp. 121-132.

Titherington, Richard H.; *History of the Spanish American War of 1898*; 1.
　　　Appleton & Co., 1900; New York.

Todleben; *Defence of Sebastopol*; Translated by William Howard Russell, 2.
　　　Tensley Bros., 1865; London.

Top News of the Week: President Nixon; National Observer, Vol. XI, No. 53, 3.
　　　Dec. 30, 1972; Silver Spring, Maryland. pg. 2.

Torpedo Armament of Various Navies; Journal of the Royal United Service 4.
　　　Institution, Vol. XXI, No. 92, Sept. 1878; London. pp. 1137-1144.

Torpedo Attack and Defence; Engineer, Vol. XLV, Apr. 5, 1878; London. 5.
　　　pp. 239-240.

Torpedo Attack on the Oberon at Stokes Bay; Engineer, Vol. XXXVIII, Aug. 14, 6.
　　　1874; London. pp. 133-134.

Torpedo Batteries; Nautical Magazine, Nov. 1865; London. pp. 614-616. 7.

Torpedo Boats and Mines; Army and Navy Journal, Vol. L, No. 13, Nov. 30, 8.
　　　1912; New York. pg. 381.

A Torpedo Detector; Engineer, Vol. XXXVII, Feb. 27, 1874; London. pg. 148. 9.

A Torpedo Detector; Van Nostrand's Engineering Magazine, Vol. X, No. 46, 10.
　　　Jun. 1874; New York. pp. 536-538.

Torpedo Experiment in Germany; Engineer, Vol. XLIV, Aug. 3, 1877; London. 11.
　　　pg. 88.

Torpedo Experiments; Army and Navy Journal, Vol. IX, No. 7, Sept. 30, 1871; 12.
　　　New York. pg. 106.

Torpedo Experiments; Engineer, Vol. XXXVIII, Aug. 21, 1874; London. pg. 152. 13.

Torpedo Experiments; Engineer, Vol. XXXVIII, Oct. 9, 1874; London. 14.
　　　pp. 271-272.

Torpedo Experiments at Newport; Army and Navy Journal, Vol. XII, No. 1, 15.
　　　Sept. 5, 1874; New York. pp. 58-59.

Torpedo Experiments at Newport; Army and Navy Journal, Vol. XIII, No. 5, 16.
　　　Sept. 11, 1875; New York. pg. 74.

1. *Torpedo Experiments in England;* Scientific American, Vol. XIV, No. 25, Jun. 16, 1866; New York. pp. 407-408.

2. *A Torpedo Explosion;* Army and Navy Journal, Vol. VIII, No. 6, Sept. 24, 1870; New York. pg. 94.

3. *Torpedo Invented by McKay;* Army and Navy Journal, Vol. III, Sept. 30, 1865; New York.

4. *Torpedo Planter;* Army and Navy Journal, Vol. XLIV, No. 18, Dec. 29, 1906; New York. pg. 479.

5. *Torpedo Planter;* Army and Navy Journal, Vol. XLIV, No. 32, Apr. 6, 1907; New York. pp. 864, 867.

6. *Torpedo Planting by the Rebels;* Army and Navy Journal, Vol. I, No. 39, May 21, 1864; New York. pg. 653.

7. *Torpedo School at Willets Point;* Army and Navy Journal, Vol. XXVI, No. 5, Sept. 29, 1888; New York. pg. 82.

8. *The Torpedo Ship;* Scientific American, Vol. XII, No. 3, Jan. 14, 1865; New York. pg. 39.

9. *Torpedo War;* Engineering, Vol. IX, Jan. 14, 1870; London. pp. 25-26.

10. *Torpedo War and Submarine Explosions;* U.S. Naval Institute Proceedings, Vol. XII, No. 2, 1886; Annapolis, Maryland. pp. 252-254.

11. *Torpedo Warfare;* Nautical Magazine, Dec. 1865; London. pp. 668-670.

12. *Torpedo Warfare;* Scientific American Supplement, Vol. XIV, No. 355, Oct. 21, 1882; New York. pp. 5659-5660.

13. *Torpedo Work in the American Civil War;* Engineering News, Vol. XXXIX, No. 16, Apr. 21, 1898; New York. pg. 249.

14. *Torpedo Work in the Civil War;* Scientific American Supplement, Vol. XLV, No. 1167; May 14, 1898; New York. pg. 18678.

15. *Torpedoes;* Army and Navy Journal, Vol. IX, No. 5, Sept. 16, 1871; New York. pg. 82.

16. *Torpedoes;* The Engineer, Vol. XVIII, Sept. 23, 1864; London. pg. 194.

Torpedoes; Engineering, Vol. I, May 18, 1866; London. pg. 333.   1.

Torpedoes; Nautical Magazine, Dec. 1865; London. pp. 663-664.   2.

Torpedoes; Van Nostrand's Engineering Magazine, Vol. I, No. 7, Jul. 1869;   3.
    New York. pp. 632-635.

Torpedoes A Century Old; Army and Navy Journal, Vol. XI, No. 27, Feb. 14,   4.
    1874; New York. pg. 426.

Torpedoes A Century Old; Engineer, Vol. XXXVII, Mar. 6, 1876; London. pg. 174. 5.

Torpedoes and Naval Mines; U.S. Naval Institute Proceedings, Vol. XLIX,   6.
    No. 6, Jun. 1923; Annapolis, Maryland pp. 1016-1018.

Torpedoes at the International Exhibition Philadelphia; Scientific American   7.
    Supplement, Vol. I, No. 25, Jun. 17, 1876; New York. pg. 388.

Torpedoes at the International Exhibition Philadelphia; Scientific American   8.
    Supplement, Vol. II, No. 31, Jul. 29, 1876; New York. pg. 479.

Torpedoes 80 Years Ago; Scientific American, Vol. LV, No. 24, Dec. 11, 1886; 9.
    New York. pp. 368-369.

Torpedoes for Defence; Army and Navy Journal, Vol. IX, No. 46, Jun. 29,   10.
    1872; New York. pp. 736-737.

Torpedoes for Defence; Army and Navy Journal, Vol. XXIV, No. 20, Dec. 11,   11.
    1886; New York. pg. 389.

Torpedoes for Defence of Harbours; Engineering, Vol. XXII, Dec. 1, 1876;   12.
    London. pp. 459-460.

Torpedoes for Defence of Harbours; Engineering, Vol. XXII, Dec. 29, 1876;   13.
    London. pp. 534-535.

Torpedoes for Our Harbors; Army and Navy Journal, Vol. XVII, No. 7,   14.
    Sept. 20, 1879; New York. pg. 121.

Torpedoes for War; Engineering, Vol. I, May 18, 1866; London. pg. 334.   15.

Torpedoes in France; Engineer, Vol. XXV, Apr. 10, 1868; London. pg. 260.   16.

Torpedoes in Turkey; Engineer, Vol. XLIV, Aug. 10, 1877; London. pg. 105.   17.

1. *Torpedoes in the United States;* Engineer, Vol. XLIII, Jan. 5, 1877; London. pg. 4.

2. *Torpedoes in War;* Army and Navy Journal, Vol. VI, Sept. 5, 1868; New York. pg. 35.

3. *Torpedoes Used by Rebels;* Scientific American, Vol. XI, No. 2, Jul. 9, 1864; New York. pg. 21.

4. *La Torpille - Mine Leon;* Le Génie Civil, Vol. LXVII, 1915; Paris. pp. 91-92.

5. *Torpilles;* Autographie du Ministère de la Marine, 1875; Paris. [Recueil factrice.]

6. Torry, John A. H.; *Minesweeping From the Air;* U.S. Naval Institute Proceedings, Vol. LXXXVII, No. 1, Jan. 1961; Annapolis, Maryland. pp. 138-140.

7. Toudouze, G. G.; *Les Nettoyeurs de la Mer;* L'Illustration, Vol. CCIV, Dec. 9, 1939; Paris. pp. 380-381.

8. Touvet, G.; *Les Mines Sous-Marines Automatiques de Contact et la Guerre de 1914-1918.* Thèse pour le doctorat presentée et soutenue le mardi 28 juin 1932. Imp. Nouvelle, 1932; Langres. 200 p. [Université de Dijon. Faculté de Droit.]

9. Townsend, A. O.; *Fulton as the Inventor of the Torpedo;* New York Times, Part 2, Jul. 18, 1915; New York. pg. 14.

10. *Trawling for Mines;* U.S. Naval Institute Proceedings, Vol. XXXV, No. 2, Jun. 1909; Annapolis, Maryland. pg. 585.

11. Trebesius, Ernst; *Triebende Seeminen und Ihre Beseitigung;* Ueberall, Vol. XVII, No. 9, 1915; Berlin. pp. 551-552.

12. Treuenfeld, R. von F.; *The Paraguayan Torpedo;* Army and Navy Journal, Vol. XII, No. 6, Sept. 19, 1874; New York. pp. 90-91.

13. *The Tricky Job of Clearing Away Those Mines;* U.S. News and World Report, Vol. LXXIV, No. 8, Feb. 19, 1973; Washington, D.C. pg. 24.

14. *Trotyl and Submarine Mines;* Journal of the U.S. Artillery, Vol. XLII, No. 42, Jul-Aug. 1914; Ft. Monroe, Virginia. pg. 375.

15. Tryckare, Tre; *The Lore of Ships;* Holt, Rinehart and Winston, 1963; New York. pp. 213-214.

Tucker, J. R., J. S. C. Gabbert, and J. W. Fothergill, Jr.; *Concept of Conducting Task Component Exercises While in Port;* Technical Report 72-2, Analytic Advisory Group, Inc., Oct. 17, 1972.   1.

Turkish Torpedoes; Nautical Magazine, Sept. 1871; London. pg. 671.   2.

Turn About; Time, Vol. XL, No. 3, Jul. 20, 1942; New York. pg. 22.   3.

Turner, John Frayn; *Service Most Silent;* George G. Harrap Co., Ltd., 1955. [The Navy's Fight Against Enemy Mines.]   4.

Über die Moralität der Submarinen Kriegfuhrung; Mittheillungen aus dem Gebiete des Sewesens, Vol. XXX, 1902; Pola. pp. 881-901.   5.

Undersea Secret; Time, Vol. XLV, No. 23, Jun. 4, 1945; New York. pg. 75.   6.

Underwater Defense; Ordnance, Vol. XX, No. 119, Mar-Apr. 1940. pp. 309-316.   7.

Unfinished Task: Moping Up; National Observer, Vol. XXI, No. 8, Feb. 24, 1973; Silver Spring, Maryland. pg. 5.   8.

United Nations (Cricket?); Time, Vol. XLIX, No. 9, Mar. 3, 1947; New York. pg. 27.   9.

U.S. Army Forces in the Pacific Survey of Japanese Coast Artillery; U.S.A.F. Pacific General Order 292 dated 27 October 1945; Report dated Feb. 1, 1946. [AFSC UF 455 J3U5]   10.

U.S. Canadian Mine Warfare Exercises Begin 6 July NATO Combined Training (Exercise Sweep Clear IV); Army, Navy, Air Force Journal, Vol. XCVI, Jul. 4, 1959; Washington, D.C. pg. 17.   11.

U.S. Destroyer Blast Blamed on Stray Mine; Evening Star and Daily News, Vol. CXX, No. 252, Friday, Sept. 8, 1972; Washington, D.C. pg. A-4.   12.

U.S. Destroyer Rocked by Blasts Off Vietnam; Washington Post, Vol. XCV, No. 227, Wednesday, Jul. 19, 1972; Washington, D.C. pg. A-18.   13.

U.S. Engineers Torpedo Manual 1898; or material of the Submarine Mining Service of the U.S. Army with a Manual for its Use in Coast Defense, 3rd Editors Press of the Battalion of Engineers, 1898.   14.

U.S. Laws, Statutes, etc., United States at Large; U.S. Government Printing Office, Vol. XXXVI, Part 2, 1911; Washington, D.C. pg. 2332.   15.

1. *U.S. Mine Work in North Sea;* Army and Navy Journal, Vol. LVI, No. 21,
    Jan. 25, 1919; New York. pp. 739-740.

2. *U.S. Mines in North Sea Removed;* Army and Navy Journal, Vol. LVII, No. 6,
    Oct. 11, 1919; New York. pg. 180.

3. *Unites States Naval Aviation 1910-1960;* NAVWEP 00-80P-1, Bureau of Naval
    Weapons, 1960; Washington, D.C.

4. *United States Naval Chronology, World War II;* CNO, Naval History Division
    Office, U.S. Government Printing Office, 1955; Washington, D.C.

5. *U.S. Navy Minesweepers and the War in South Viet Nam;* Navy, Vol. V, No. 8,
    Aug. 1962; Washington, D.C. pg. 20.

6. *U.S. Resumes Mining of Haiphong's Harbor: Above 20th Parallel;* Evening
    Star, Vol. CXX, No. 353, Monday, Dec. 18, 1972; Washington, D.C.
    pp. A-1, A-6.

7. U.S. School of Submarine Defense; *Annual Report of the Commandant 1902-1903;*
    Oct. 15, 1903; Ft. Totten, New York.

8. *U.S. Still Reseeds Mine Fields in N. Viet Harbors;* Washington Evening Star,
    Vol. CXX, No. 342, Thursday, Dec. 7, 1971; Washington, D.C. pg. A-5.

9. U.S. Treaties, etc.; *Laying of Automatic Submarine Contact Mines (Hague VIII);*
    U.S. Treaties and Other International Agreements, Vol. I, U.S. Government
    Printing Office, 1968; Washington, D.C. pg. 669.

10. Usborne, Cecil Vivian; *Blast and Counterblast;* John Murray, 1935; London.

11. *Use of Mines at Sea;* Illustrated London News, Vol. CXLVIII, 1916; London.
    pg. 438.

12. *Use of Mines in Russo-Japanese War;* U.S. Naval Institute Proceedings; Vol.
    XL, No. 5, Sept-Oct. 1914; Annapolis, Maryland. pp. 1507-1508.

13. *The Use of the Mine;* United Service Gazette, Vol. CLXIII, Oct. 8, 1914;
    London. pp. 269-270.

14. Vance, Willis R.; *Observation Firing of Submarine Mines;* Journal of the U.S.
    Artillery, Vol. XXXI, No. 2, Mar-Apr. 1909; Ft. Monroe, Virginia.
    pp. 144-150.

Van der Veer, Norman R.; *Mining Operations in the War;* U.S. Naval Institute   1.
 Proceedings, Vol. XLV, No. 11, Nov. 1919; Annapolis, Maryland. pp.
 1857-1865.

Van Duzer, E. M.; *Diesel Sweepers Lead Attack;* Motorship, Vol. XXIX, No. 9,   2.
 Sept. 1944; Seattle and New York. pp. 776-778.

Van Nostrand, A. D.; *Minesweepers;* U.S. Naval Institute Proceedings, Vol.   3.
 LXXII, No. 4, Apr. 1946; Annapolis, Maryland. pp. 505-509.

Vergos, Lt de vaisseau; *Cours de Substances Explosives;* Republique Française,   4.
 Marine Nationale, Exemplaire No. 15, Imprimerie Nationale, 1899; Paris.
 106 p. [Confidentiel.]

The Versatile Minesweeper; Navy, Vol. V, No. 8, Aug. 1942; Washington, D.C.   5.
 pp. 18-19.

Veth, Kenneth L.; *Iron Men in Wooden Ships of (Pacific) Mine Force;* Army,   6.
 Navy, Air Force Journal and Register, Vol. C, Jun. 6, 1963; Washington,
 D.C. pp. 18-19.

Veth, Kenneth L.; *Mine Warfare; State of the Art;* Sperryscope, Vol. XVI,   7.
 No. 3, 1962; Brooklyn, New York. pp. 12-15.

Vetkin, K. S.; *After the Sweepers - Clear Sailing;* Morskoy Sbornik, No. 6,   8.
 Jun. 1971. pp. 84-90. [Translation; Original pp. 34-43.]

(Vickers-Elia); *Submarine Mines (Automatic Submarine Mines);* Brassey's   9.
 Naval Annual, 1912; London. pp. 313-316.

Villers, G.; *Devotion to Duty Above All;* U.S. Naval Institute Proceedings,   10.
 Vol. LXXXIV, No. 10, Oct. 1958; Annapolis, Maryland. pg. 82.

Virginskii, Viktor Semanovich; *Robert Fulton 1765-1815;* 'Hayka', 1965;   11.
 Moscow.

Vogel, Bertram; *The Great Strangling of Japan;* U.S. Naval Institute Pro-   12.
 ceedings, Vol. LXII, No. 11, Nov. 1947; Annapolis, Maryland. pp.
 1304-1309.

Vohtz, J. A.; *Det Passive Sømine Materiel Undar-bejdet til Brug ved*   13.
 Søminekarpset; Orlogsvaerftet, 1893.

Von der Porten, Edward P.; *The German Navy in World War II;* Thomas Y.   14.
 Crowell Co., 1969; New York.

1. Vrangel, Ferdinand F.; *Vice-Admiral S. O. Makarov: A Biographical Sketch;* Maritime General Staff, 2 volumes, 1911, 1913; St. Petersburg.

2. (Vulcan); Naval Chronicle, Vol. XXI, Joyce Gold, 1809; London. pp. 452-453.

3. Wade, H. A. L. H.; *Two Destroyers (British, Damaged in 1946 Off Corfu);* Journal of the Royal United Service Institution, Vol. XCIX, No. 593, Feb. 1954; London. pg. 62.

4. Wadlow, E. C.; *Mechanical Engineering Aspects of Naval Mining;* Engineer, Vol. CLXXV, Feb. 13, 1948; London. pp. 165-166.

5. Wadlow, E. C.; *Mechanical Engineering Aspects of Naval Mining;* Proceedings of the Institute of Mechanical Engineers, Vol. CLX, No. 1, Jun. 1940; London. pp. 23-32, 44-51.

6. Wages, C. J., Jr.; *Mines ... the Weapons that Wait;* U.S. Naval Institute Proceedings, Vol. LXXXVIII, No. 5, May 1962; Annapolis, Maryland. pp. 102-114.

7. Wagner, Fredrick; *Submarine Fighter of the American Revolution;* Dodd, Mead, 1963; New York.

8. Wainwright, Lawrence; *The Mine as a Tool of Limited War;* U.S. Naval Institute Proceedings, Vol. XCIII, No. 6, Jun. 1967; Annapolis, Maryland. pg. 105.

9. Walke, Rear Admiral H.; *Naval Scenes and Reminiscences ...;* F. R. Reed, 1877; New York. xii, 480 p.

10. Walker, J. Bernard; *Landsman's Log Aboard the United States Destroyer "Patterson" - II;* Scientific American, Vol. CVII, No. 16, Oct. 19, 1912; New York. pp. 328-329.

11. Walker, J. Bernard; *Our Shortage of Scouts, Torpedoes and Mines;* Scientific American, Vol. CX, No. 14, Apr. 4, 1914; New York. pp. 281-282.

12. Walker, J. Bernard; *The Submarine Problem: Closing the North Sea with a Bomb-Curtain;* Scientific American, Vol. CXVI, No. 25, Jun. 23, 1917; New York. pp. 616-617.

13. Walker, Sydney F.; *Submarine Engineering: All About Work Underwater Diving, Submarines, Torpedoes, Cables, Dredging, etc.;* C. Arthur Pearson Ltd., 1914; London. [Chapter IV, pp. 37-42.]

Walsh, W. T.; *Gunning for Lost Mines*; Illustrated World, Vol. XXXI, Apr.   1.
1919; Chicago. pp. 186-189.

Wandevelde, Capt Lt de Marine J. A.; *Rapport de la Commission des Torpilles*   2.
Assemblée à Brielle, pour Faire des Expériences sur le Passage des
Bateaux au-dessus des Torpilles; Traduit du Neerlandais par M. W.
Kamps, J. Corréard, 1869; Paris. 54 p.

Wang Shi Fu; *Naval Strategy in the Sino-Japanese War*; U.S. Naval Institute   3.
Proceedings, Vol. LXVII, No. 7, Jul. 1941; Annapolis, Maryland. pp.
991-998.

War; The Times Literary Supplement, No. 2841, Aug. 10, 1956; London. pg. 480.   4.

War; The Times Literary Supplement, No. 2956, Oct. 24, 1958; London. pg. 615.   5.

The War at Sea; Newsweek, Vol. XV, No. 2, Jan. 8, 1940; New York. pp. 21-22.   6.

War at Sea: Death for the Magpie; Time, Vol. LVI, No. 16, Oct. 16, 1950;   7.
New York. pg. 27.

The War at Sea: Going In; Time, Vol. LVI, No. 18, Oct. 30, 1950; New York.   8.
pg. 35.

War at Sea: Mines Ahead; Time, Vol. LVIII, No. 17, Oct. 22, 1951; New York.   9.
pg. 27.

Warneck, P. A.; *The Baltic Straits During the First World War*; The Naval   10.
Records, Vol. X, 1952; Moscow. pp. 10-16.

Warner, Oscar C.; *The German Naval Offensive and Defensive Mine*; Journal of   11.
the U.S. Artillery, Vol. XLIII, May-Jun. 1915; Ft. Monroe, Virginia.
pp. 336-339.

Warren, Harris Gaylord; *Paraguay, An Informal History*; University of   12.
Oklahoma Press, 1949; Norman, Oklahoma.

Warshofsky, Fred; *War Under the Waves*; Pyramid Books, R-791, 1962; New York. 13.

Washburn, S.; *Floating Mines in Naval Warfare*; Outlook, Vol. LXXXVI, Jun. 8, 14.
1907; Chicago. pp. 281-286.

1. Washington, John; Report on Visit to Kronstadt, September 1853; Napier
        Papers. Additional Manuscripts 40023, Vol. VI, dated Oct. 1, 1853
        (with Sept. 23 crossed out). [British Museum. Literature Reading Room.]

2. Webster, Charles and Noble Frankland; *History of the Second World War.*
        The Strategic Air Offensive Against Germany, 1939-1945; Her Majesty's
        Stationery Office, Vols. 1-4, 1961; London.

3. Welles, Gideon; *Diary of Gideon Welles, Secretary of the Navy;* 3 volumes:
        Vol. 1, 1861-Mar. 30, 1864, 549 p.; Vol. 2, Apr. 1, 1864-Dec. 31, 1866,
        653 p.; Vol. 3, Jan. 1, 1867-Jun. 6, 1869, 672 p.; Houghton, 1911:
        Boston.

4. Welling, William G.; *Minelaying By Aircraft;* Ordnance, Vol. XL, No. 219,
        Nov-Dec. 1956; Washington, D.C. pp. 415-418.

5. Welliver, J. C.; *A Minesweeping Cruise in the Submarine Zone;* Munsey, Vol.
        LXI, No. 2, Jul. 1917; New York. pp. 241-243.

6. Wells, L.; *England's Death and Glory Boys Who Deactivate Unexploded Bombs;*
        Reader's Digest, Vol. XXI, No. 245, Sept. 1942; Pleasantville, New York.
        pp. 39-42.

7. Wester-Wemyss, R. E. W.; *The Navy in the Dardanelles Campaign;* Hodder and
        Stoughton, 1924; London.

8. Wettern, Desmond; *All Quiet for the Oysters;* U.S. Naval Institute Proceedings,
        Vol. LXXXV, No. 1, Jan. 1959; Annapolis, Maryland. pp. 138-139.

9. Wettern, Desmond; *Royal Navy Operates Mine-Hunting Sonar;* U.S. Naval Institute
        Proceedings, Vol. LXXXIX, No. 3, Mar. 1963; Annapolis, Maryland.
        pp. 144-145.

10. (Wheeler, J. B.); *New Publications;* Army and Navy Journal, Vol. XXI, No. 36,
        Apr. 5, 1884; New York. pg. 744. [Book review.]

11. *When A Periscope Does Not Mean A Submarine;* Scientific American, Vol. CXVIII,
        Apr. 27, 1918; New York. pg. 383.

12. White, Fedoroff D.; *Survival Through War and Revolution in Russia;* University
        of Pennsylvania Press, 1939; Philadelphia.

13. White, Joe W.; *My Minesweeping Experiences;* Wide World Magazine, Vol. XLIII,
        1919; London. pp. 421-427.

White, W. L.; *Day of Sweeping Mines Off Dover*; Reader's Digest, Vol.   1.
    XXXVIII, No. 227, Mar. 1941; Pleasantville, New York. pp. 95-99.

Whiteman, Marjorie M.; *Digest of International Law*; U.S. Government   2.
    Printing Office, Vol. X, 1968; Washington, D.C. pg. 681.

Why Hanoi Realized It "Cannot Hope to Win"; U.S. News and World Report,   3.
    Vol. LXXIII, No. 19, Nov. 6, 1972; Washington, D.C. pp. 17-19.

Why Vietnam War Drags On; U.S. News and World Report, Vol. LXXIV, No. 1,   4.
    Jan. 1, 1973; Washington, D.C. pp. 9-12.

Wie Konnen Kriegschiffe gegen die Wirkungen von Torpedoes und Minen   5.
    Geschutzt Werden; Ueberall, Vol. II, No. 18, 1904; Berlin. pp. 1-2.

Wildrick, Meade; *Effect Upon Measures for Coast Defense of the Develop-*   6.
    ment of Submarine and Aerial Attacks; Journal of the U.S. Artillery,
    Vol. XLV, No. 2, Mar-Apr. 1916; Ft. Monroe, Virginia. pp. 145-180.

Williams, Frances L.; *Mathew Fontaine Maury: Scientist of the Sea*;   7.
    Rutgers University Press, 1963; New Brunswick.

William's System of Coast Defense by Electrical Torpedoes; Scientific   8.
    American Supplement, Vol. XVIII, No. 467, Dec. 13, 1884; New York.
    pp. 7451-7452.

Wilmot, Chester (Reginald William Winchester Wilmot); *The Struggle for*   9.
    Europe; Collins, 1952; London.

Wilson, George C.; *Intensive Bombing of North Resumed: U.S. Planes Hit*   10.
    Hanoi Haiphong; Washington Post, Vol. XCVI, No. 14, Tuesday,
    Dec. 19, 1972; Washington, D.C. pp. A-1, A-19.

Wilson, George C.; *New Bomb Eyed for Vietnam Use*; U.S. Naval Institute   11.
    Proceedings, Vol. XCIII, No. 12, Dec. 1967; Annapolis, Maryland.
    pg. 146.

Wilson, George C.; *N. Viet Harbors to Be Swept: U.S. Ready to Clear*   12.
    Mines; Washington, Post, Vol. XCV, No. 341, Friday, Nov. 10, 1971;
    Washington, D.C. pp. A-1, A-19.

Wilson, George Grafton (Compiler); *International Law and Discussions 1914*;   13.
    U.S. Government Printing Office, 1915; Washington, D.C.

1. Wilson, H. W.; *Battleships in Action;* Little, Brown, 2 volumes, 1926; Boston.

2. Wilson, H. W.; *The Downfall of Spain Naval History of the Spanish American War;* Sampson Low, Marston, Co., 1900; London.

3. Winn, Frank L.; *Torpedoes and Submarine Mines;* United Service, New Series, Vol. VIII, No. 5, Nov. 1892; Philadelphia. pp. 464-475.

4. Winterhalder, Theodore; *Die Osterreichisch-Ungarische Kriegsmarine im Weltkrieg;* J. F. Lehmann, 1921; Munich. [Pamphlet.]

5. *Wire Drag Work of the U.S. Coast and Geodetic Survey;* Engineering News, Vol. LXIV, No. 22, Dec. 1, 1910; New York. pp. 88-89.

6. Wisser, John P.; *The Strategy and Tactics of the Russo-Japanese War;* United Service, Vol. VIII, 3rd Series, 12th Paper, Sept. 1905; New York. pp. 201-213.

7. Wisser, John P.; *The Tactics of Coast Defense;* Hudson-Kimberly Publishing Co., 1902; Kansas City, Missouri.

8. Wisser, John P.; *War Lessons for the Coast Artillery;* Journal of the U.S. Artillery, Vol. XXII, No. 3, Nov-Dec. 1904; Ft. Monroe, Virginia. pp. 267-269.

9. *With a Hoop and a Whoop;* Time, Vol. XLI, No. 4, Jan. 25, 1943; New York. pg. 26.

10. Wittmer, R.; *Die Torpedowaffe;* Königliche Fredrich-Wilhelms Universität-Institut für Meerskunde, 1909; Berlin.

11. Wolfram, E.; *Minen im Kattegatt;* Marine Rundschau, Vol. XXXIX, 1934; Berlin. pp. 493-500.

12. Wood, A. B.; *From Board of Invention and Research to Royal Naval Scientific Service;* Journal of the Royal Naval Scientific Service, Vol. XX, No. 4, Jul. 1965. [Memorial Number to A. B. Wood, O.B.E., D.Sc.]

13. Wood, Walter; *Mine Sweeping;* Navy and Army Illustrated, Vol. I, London, New Series, 1914. pp. 147-148.

14. Woodbury, David Oakes; *What the Citizen Should Know About Submarine Warfare;* Norton, 1942; New York.

15. *Wooden-Built Motor Minesweeper in Clearing Channels of Acoustic and Magnetic Mines;* Illustrated London News, Vol. CCVIII, Jun. 8, 1946; London. pg. 626.

Wooden Ships Have Work to Do; Sea Power, Vol. II, No. 2, Feb. 1942;   1.
   Washington, D.C. pp. 26-27.

Woodhouse, Henry; *Text Book of Naval Aeronautics;* The Century Co., 1917;   2.
   New York.

Woodward, David; *The High Seas Fleet, 1917-1918;* Journal of the Royal   3.
   United Service Institution, Vol. CXIII, No. 651, Aug. 1968;
   London. pp. 244-250.

Woodward, David; *Munity at Wilhelmshaven;* History Today, Vol. XVIII, No.   4.
   11, Nov. 1968; London. pp. 779-785.

Woodward, David; *The Russians at Sea;* Praeger, 1966; New York.   5.

Work of American Mine Sweepers; Current History Magazine, New York Times,   6.
   Vol. XI, Part 2, Jan. 1920; New York. pp. 68-69.

Work of the Army Engineers; Army and Navy Journal, Vol. XXXV, No. 28,   7.
   Mar. 12, 1898; New York. pp. 512-513.

Wulff, Olaf Richard (Vizeadmiral); *Die Österreichisch-Ungarische*   8.
   Donauflotille im Weltkrieg, 1914-1918; W. Braumüller, 1934; Vienna.

"X"; *Der Minenkrieg;* Ueberall, Vol. VIII, No. 32, 1906; Berlin. pp. 380-381.   9.

Yarham, E. R.; *The War in Nazi Ore Routes;* Nautical Magazine, Vol. CLII,   10.
   Sept. 1944. pp. 164-167.

Yonge, Charles Duke; *The History of the British Navy ...;* R. Bentley, 2nd   11.
   Edition, 3 volumes, 1866; London.

Zakharov, S. E.; *A History of the Art of Naval Warfare;* National Technical   12.
   Information Service, U.S. Department of Commerce, JPRS-52287, Jan. 29,
   1971. [Translation.]

Index of Authors
on
Naval Mine Warfare

Code

(A)	Author
(C)	Collaborator
(E)	Editor
(NP)	Nom de Plume
(P)	Publisher
(T)	Translator

Name(s)	Page(s)
Abbot, Henry Larcom	1
Abel, F. A.	1
Adams, William T.	1
Albert, Prince de Monaco	2
Albion, Robert Greenhalgh (C), see Pope, Jennie Barnes	2
Alden, Carroll S. (C), see Westcott, Allan	2
Alden, John D.	2
Alexander, Ray	2
Alexinsky, Gregor	2
Alger, Philip R., Jr.	2
Allan, Westcott (C), see Alden, Carroll S.	2
Alliman, A. L.	3
Allison, R.	3
Ammen, Daniel	3
Anderson, Bern	3
Anderson, Frank	3
Anderson, Jack	3
Anderson, Jane (C), see Bruce, Gordon	3
Andrews, A. F. (C), see Hoang, P. R.	3
Anglrer, J. F. (C), see Lankford, B. W.	54
Ansel, Walter	4
Anthony, Irvin	4
Appleton, C.	4
Armagnoe, A. P.	4
Armstrong, Lt G. E. (A), see Robinson, Charles Napier (E)	88
Armstrong, James	4
Ashbrook, Allan Withes	4

Name(s)	Page(s)
Ashley, L. R. N.	5
Ashton, George	5
Aspinall-Oglander, Cecil F.	5
Assmann, Kurt	5
Aston, George G.	5
Audic, M. (NP)	5
Auphan, Paul (C), see Mordal, Jacques	5
Avery, Ray L.	6
Avezathe, B.	6
Bagby, Oliver W.	6
Baggott, A. J. (C), see Fawcett, C. H.	6
Baird, C. W.	6
Baird, George Washington	7
Baker, H. G.	7
Barber, Francis Morgan	7
Barclay, Thomas	7
Barnard, Henry	7
Barnes, John Sanford	7
Barnes, Lt Commander	7
Barrow, John	7
"Bartimeus" (NP)	7
Bartlett, C. J.	8
Bates, L. M. (C), see Grosvenor, Joan	41
Bates, P. L.	8
Baugh, Barney	8
Beals, Victor	8
Beahler, W. H. (T)	8
Belden, Frank A. (C), see Haven, Charles T.	44
Belknap, Reginald Rowan	8
Bell, Archibald Colquhoun	9
Bell, L.	9
Bellet, Daniel	9
Benjamin, Dick (C), see Gravat, J. J.	9
Benjamin, P.	9
Bent, E. D.	9
Benton, Elbert J.	9
Berg, Ernst	9
Bernay, Henri	9
Bernotti, R.	9
Bertin, E.	10
Bethell, John	10
Betts, J. A.	10
Beucker, A Lt de vaisseau (C), see Ellis, A. G.	31

-121-

Name(s)	Page(s)
Bird, A.	10
Bishop, Farnhorn	10
Bizot, M. (T), see Seuter, M. (A)	10
Blackman, Raymond V. B. (E)	10
Blakeslee, H. W.	10
Blic, Lt de vaisseau de	10
Bock, Ingeborg	10
Boesen, Victor (C), see Karneke, Joseph Sidney	52
Boggs, H. Glenn	10
Boling, Gerald	11
Bolton, Reginald	11
Bone, M.	11
Bonny, A. D.	11
Bourgois, Simeon	11
Bowen, Frank C.	11
Bowers, John V.	11
Bowler, Roland Tomlin E. (E)	11, 12
Boyd, Willim B. (C), see Rowland, Buford	89
Boynton, Charles B.	12
Bradsher, Henry S.	12
Bragadin, Marc' Antonio	12
Brainard, Alfred P.	12
Bramah, F., Jr.	12
Brasseur, Pierre	12
Brassey (P)	12
Bravetta, E.	12
Breyer, Siegfried	12
Bristol, Jack A.	12
Brodie, Bernard	13
Brookes, Ewart	13
Brookfield, S. J.	13
Brou, Willy Charles	13
Brown, William Baker	13
Bruce, Gordon (C), see Anderson, Jane	3
La Bruyere, R.	13
Bryant, Samuel W.	13
Buchard, H.	13
Buckey, Mervyn Chandos	13
Bucknill, John Townsend	13, 14
Bultman, H. F. E.	14
Bunker, Paul D.	14
Burgoyne, A. H.	15
Busch, Fritz Otto	15
Bushnell, David	15

Name(s)	Page(s)
Butterworth, A.	15
Cadiat, Ernest (C), see Lediau, Alfred	55
Cagle, Malcolm W. (C), see Manson, Frank A.	15
Callender, M. L.	15
Callwell, C. E.	15
Calme, Barney	15
Calme, Byron E.	16
Campbell, John	16
Canfield, Eugene B.	16
Capehart, E. E.	16
Carter, W. R. (C), see Duvall, E. E.	16
Cassell (P)	16
Cate, James Lea (C), see Craven, Wesley Frank	22
Catlin, George L.	16
Chadwick, French Ensor	16
Champlin, G. F.	16
Chardonneau, F.	16, 17
"Chasseur" (NP)	17
Chatterton, E. Keble	17
Chatterton, Howard A.	17
Chinn, W. K. (C), see Rogers, T. H.	88
Churchill, W. S.	17
Clark, Frank S.	18
Clark, Thomas	18
Clark, William Bell (E)	18
Clarke, G. S.	18, 19
Clarke, Lt-Col Sir Georges S. (C), see Thursfield, James R.	19
"Claudeville" (NP)	19
Cleator, P. E.	19
Clouet, Alain	19
Clowes, William Laird	19
Cluverius, Wat Tyler	19
Colden, Codwalder D.	19
Coleman, Frank	19
Coletta, Paola E.	20
Colombos, John Constantine	20
Comey, Robert W.	20
Condon, J. F. (C), see Koslow, H. M.	20
Conness, Leland	20
Conti, A.	20
Cook, Gilbert	21
Cooper, James Fenimore	21
Corbett, Julian S.	21

Names(s)	Page(s)
Corbin, Diana F. M.	21
Cowie, John S.	21, 22
Craighill, W. E.	22
Cranford, L. Cope	22
Craven, Wesley Frank (C), see Cate, James Lea	22
Crouse, George M.	22
Crowley, R. O.	22
Crowther, James Gerald (C), see Wheddington, R.	23
Cummings, Joseph D.	23
Cunningham, Andrew B.	23
Cutbush, James	23
Dahl, E. M.	23
Dallin, David J.	23
Daly, Robert W.	23
Daniel, C. S.	23
Daniels, Josephus	23
Dauch, Lt de vaisseau	24
Daudenart, Major d'État-Major L. G.	24
Davelny, Contre-Amiral	24
Davelny, Rene	24
Davidson, Hunter	24
Davies, G.	24
Davis, Ewing O.	24
Davis, G. H.	24
Davis, J. H.	24
Davis, Noel (E)	25
Davis, Richmond P.	25
Dawson, W.	25
De Gouy, Jean Baptiste (C), see Mathieu, C. R.	26
DeGreene, K. B.	26
DeKerc'hoat, L.	26
Delafield, R.	26
Delegue, Cap de Frégate	26
Delpeuch, Lt de vaisseau Maurice	26
Demigny, A.	26
Dempewolff, Richard F.	26
Denisov, B. A. (C), see Goncharov, L. G.	40
Denlinger, Sutherland (C), see Gary, Charles P.	26
Devize (ou Deveze), A.	27
Dibos, M.	27
Dickinson, H. W.	27
Dickinson, J. H. (T), see Klado, Nicholas L. (A), Marchant, F. P. (T)	53
Didelot, Carl	27

Name(s)	Page(s)
Doflein, M.	28
Dohna-Schlodien, Burgrave Nicolas zu	28
Dollo, A.	28
Dommett, William Erskine	28
Domville-Fife, Charles W.	28
Dooly, William G.	28
Dorling, H. Taprell ("Taffrail")	28, 103
Doty	28
Douglas, Harold G. (C), see Snow, Chester R.	96
Duke, Irving Terri	29
Dukers, Capitaine de Frégate	29
Duncan, Robert C.	29
Dunn, John M.	29
Dunsany, Admiral Lord	29
Durassier, Edward	29
Duvall, E. E. (C), see Carter, W. R.	16
Dzienisiewicz, Cap de Corvette	29
Earp, G. Butler	29
"E.B." (NP)	29
Edmonds, James E.	29
Edwards, Harry William	30
Edwards, Kenneth	30
Egan, Richard	30
von Ehrenkrook, Fredrich (Lt de vaisseau L. von)	30
Eisner, Will	30
Elia, Giovanni Emanuelle	31
Eliot, G. F.	31
Eller, Ernest MacNeill	31
Ellicott, John M.	31
Ellis, A. G. (C), see Beucker, A Lt de vaisseau	31
Ellis, William S.	31
Englehart, Alva F.	31
Esper, George	32
Evans, K. G.	32
Evelegh, M. H.	32
Fane, Francis D. (C), see Moore, Don	33
Fane, Robert (NP)	33
Farnworth, Ellis (T), see Machiavelli, Niccolo (A)	59
Fawcett, C. H. (C), see Baggott, A. J.	6
Fayles, Charles Ernest	33
Fein, Maier O. (C), see Masterson, Harold	33

Name(s)	Page(s)
Ferguson, J. N.	33
Ferraby, H. C.	33
Ferrand, M. C.	33
Ferris, G. T.	33
Field, James A.	33
Fisher, John A.	34
Fisher, Richard	34
Fiske, Bradley Allen	34
Fleming, Peter	34
Fontin, Paul	35
Forbes, C.	35
Ford, J. H.	35
Forrest, G.	35
Forshell, Hans	35
Fothergill, J. W., Jr. (C), see Gabbert, J. S. C. or Tucker, J. R.	109
Fournier, F. E.	35
Le Franc, A.	35
Francis, D.	35
Frankland, Noble (C), see Webster, Charles	114
Fraser, I. S.	35
Frazier, R. W.	36
Freeman, Lewis R.	36
French, W. F.	36
Frothingham, Thomas G.	36
Fuller, J. F. C.	36
Fulton, Robert	36, 37
Fuqua, S. O.	37
Fyfe, Herbert C.	37
Gabbert, J. S. C. (C), see Tucker, J. R. or Fothergill, J. W., Jr.	109
Gakauma, G.	37
Gale, Benjamin	37
Galland, Adolf (A), see Savill, Merryn (T)	37
Galvin, William M.	37
Garcia, R. C.	37
Garner, J. W.	37
Garvin, Richard L.	38
Gary, Charles P. (C), see Denlinger, Sutherland	26
Gaunt, Richard H.	38
Gawn, R. W. L.	38
Getler, Michael	39
de Ghappedelaine, G. (T)	39
Gibson, Charles R.	39
Gibson, R. H. (C), see Pendergast, Maurice	39
Gillett, J. K. (C), see Schoute-Vanneck, C. A.	39
Gilmore, Arthur H.	40
Ginocchetti, Angelo	40

Name(s)	Page(s)
Gisborne, F. G.	40
Goncharov, L. G. (C), see Denisov, B. A.	40
Goodeve, C. F.	40
Goodrich, Casper Frederick	40
Gordon, C. V.	40
Gorman, Ron (C), see Rodgers, Hamp	88
Gosse, Joseph	40
Graf, George (C)	40
Graf, H.	40
Grant, Robert M.	41
Grasset, Albert	41
Gravat, J. J. (C), see Benjamin, Dick	9
Gray, Edwyn	41
Gray, Elisha	41
Green, Fitzhugh	41
Greenough, Ernest A.	41
Grenfell, Russell	41
Gretton, Peter W.	41
Griswold, Charles	41
Grivel, R.	41
Grosjean, M.	41
Grosvenor, Joan (C), see Bates, L. M.	41
Groves, Donald	42
Guichard, Louis (A), see Turner, Christoper R. (T)	42
Guierre, A.	42
Gwynne, Alban Lewis	42
Hahn, Fritz	42
Hailey, Foster (C), see Milton, Lancelot	42
Haldane, J. B. S.	42
Hall, Charles H.	42
Hall, Cyril	42
Hall, J. A.	42
Halsey, F. W.	42
Hamilton, John Randolph	43
Hampshire, A. Cecil	43
von Handel-Mazzetti, P. A.	43
Hanks, Carlos C.	43
Hansards (P)	43
Hansen, H.	43
Hare, Robert	43, 44
Harriman, J. E.	44
Harris, Robert Hastings	44
Hartley, A. B.	44
Hartmann, Gregory K.	44
Harvey, John	44

Name(s)	Page(s)
Hauck, Russel	44
Haven, Charles T. (C), see Belden, Frank A.	44
Hay, D. (C), see Postan, M. M. or Scott, J. D.	83
Hayes, John D. (E)	44
"H.B." (NP)	44
Heald, Joseph F.	44
Heinl, Robert D., Jr.	45
Hennebert, Eugene	45
Henningsen, Charles Fredrick	45
Herrick, Robert Waring	45
Hershey, Amos S.	45
Hessler, William H.	45
Higgins, Alexander Pearce	45
Hine, Al	45
Hinkamp, Clarence Nelson	46
Hirshberg, N. (C), see Piehler, D.	81
Hoang, P. R. (C), see Andrews, A. F.	3
Hoblitzell, James J.	46
Hobson, Richmond P.	46
Hoehling, Adolph A.	46
Holman, H. R.	46
Holmes, Nathaniel J.	46
Holmes, Wilfred J.	47
Holt, David H.	47
Hornsnaill, W. O.	47
Houllevigue, L.	47
Hubbard, J. C.	47
Hubbard, Miles H.	47
Hubbe, M.	47
Huber, James	47
Huet, C.	47
Hughes, Hobart	48
Hull, E. W. Seabrook	48
Hurd, A. S.	48
Hurd, Archibald	48
Hurwitt, Albert	48
Husnu, Kurt	48
Hyman, Jerome	48
Ipsen, P.	49
Irwin, A. E.	49

Name(s)	Page(s)
"J.A.K." (NP)	49
Jane, Fredrich T.	49, 50
Jaques, William Henry	50
Jeffers, William N.	50
Jellicoe, John Rushworth	50
Jervois, W. F. Drummond	50
Johns, A. W.	50
Johnson, Bruce H.	50
Johnson, Ellis A. (C), see Katcher, David A.	50
Johnson, John	50
Johnston, Oswald	50
Jones, C. B.	51
Jones, Henry L.	51
Jones, H. Robert	51
Jones, J. M.	51
Jones, J. William	51
Jones, Virgil Carrington	51
Jose, A. W.	51
Julicher, Peter J.	51
Juul, C.	51
Kalmpffert, W.	52
Kannengiesser, Hans Pasha	52
Karig, Walter Cagle (C), see Manson, F. A.	52
Karl, R. L. (C), see Thorton, J. H., Jr.	52
Karneke, Joseph Sidney (C), see Boesen, Victor	52
Katcher, David A. (C), see Johnson, Ellis A.	50
Kauffman, Draper L.	52
Kay, Howard N.	52
Keate, E. M. (C), see Tunstall, W. C. B.	52
Kekewich, Piers K. (A), see Larghezza, G. (T)	52
Kelly, H. W. K.	52
Kelly, Orr	52
Kiep, U. H. A.	52
Kim, Song Mo	52
King, Cecil	53
King, Ernest J. (C), see Whitehill, Walter M.	53
King, William Rice	53
Kingsford, H.	53
Kinney, Sheldon	53
Kirby, S. Woodburn	53
Kirk, John (C), see Young, Robert, Jr.	53
Klado, Nicholas L. (A), see Dickinson, J. H. (T) or Marchant, F. P. (T)	53
Knapp, H. S.	53
Knox, Dudley W.	53
Knox, Thomas W.	53

Name(s)	Page(s)
Kolbenschlog, George R.	53
Korotkin, I. M.	54
Koslow, H. M. (C), see Condon, J. F.	20
Krafft, Herman F. (C), see Norris, Walter B.	54
Kraus, J. H.	54
Kvam, Kare E.	54
Lake, Simon	54
Lamont, R. R.	54
Langmaid, Kenneth J. R.	54
Lankford, B. W. (C), see Anglrer, J. R.	54
Lankford, B. W. (C), see Pinto, J. E.	54
Larghezza, G. (T), see Kekewich, Piers K. (A)	52
Larghezza, G.	54
Laubeuf, Alfred Maxime (C), see Stroh, Henri	54
Lauth, J.	54
Lawrence, T. J.	55
Lebedskoi, G. M.	55
Ledieu, Alfred (C), see Cadiat, Ernest	55
Ledig, Gerhard	55
Legrand, J.	55
Lepotier, Capt de Vais A.	55
Levie, Howard S.	55
Lewis, Charles Lee	55, 56
Lewis, Michael	56
Liman von Sanders, O. V. K.	56
Lincoln, Fredman Ashe	56
Lissak, Ormond M.	56
Lloyd, Christopher (E)	56
Lochner, R.	57
Lockroy, Edouard	57
Lohman, Walter	57
Lohr, Carl A.	57
Long, A.	57
Loosbrock, J. F.	57
Lorey, Hermann	57
Lossing, Benson J.	57
Lott, Arnold S.	57, 58
Lott, Davis Newton	58
Loughton, L. G.	58
Loukine, Alexandre	58
Low, Archibald Montgomery	58
Lukin, A. P.	58
Lundeberg, Philip K.	58
Lupinacci, Pier Fillippo (C), see Tognelli, Vittorio	58
Lusar, Rudolph	58
"M" (NP)	58
"M.A.W." (NP)	59

Name(s)	Page(s)
McClintock, Robert	59
McClung, Frank	59
MacDonald, H.	59
McEarthron, Ellsworth D.	59
McElgin, Hugh J. B.	59
McEntee, Gerard Lindsley	59
McGrath, Thomas D.	59
MacGregor, Edar John	59
Machiavelli, Niccolo (A), see Farnworth, Ellis (T)	59
McIlwraith, Charles G.	60
MacIntrye, Donald	60
McKay, Robert F.	60
McLachlon, Bruce	60
Maclean, Alstair S.	60
McMurtrie, Francis	60
de Maconge, J. L.	60
Magruder, T. P.	61
Mahan, Alfred Thayer	61
Malcolm, E. D.	61
Mallot, Robert	61
Manfroni, C.	61
Mann, C. F. A.	61, 62
Mannix, D. P.	62
Manson, Frank A. (C), see Cagle, Malcolm W.	15
Manson, Frank A. (C), see Karig, Walter Cagle	52
Marchant, F. P. (T), see Klado, Nicholas L. (A), or Dickinson, J. H. (T)	53
Marder, Arthur J.	62
Marder, Murrey	62
Marks, E.	62
Marshall, H. W. S.	63
Martienssen, Anthony	63
Martin, L. W.	63
Masterman, J. C.	63
Masterson, Harold (C), see Fein, Maier O.	33
Mathieu, C. R. (C), see De Gouy, Jean Baptiste	26
Maury, Matthew Fontaine	63
Maury, Richard Lancelot	63
Mayo, Claude Banks	63
Meacham, James Alfred	63
Madlicott, W. N.	64
Meister, Jurg	64
de Mello Tamborim, A. J.	64
Mercer, David D.	64
Merrifield, C. W.	64, 65
Michaels, R.	65
Michel, N. B.	65
Mielichhofer, Sigmund	65

Name(s)	Page(s)
Milbury, C. E.	65
Miles, A. H.	65
Miller, Henry A.	66
Miller, Michael	66
Miller, Richards T.	66
Miller, N. J.	66
Millholland, Ray	66
Milligan, John D.	66
Milton, Lancelot (C), see Hailey, Foster	42
Mintzer, Leonard Murney	72
Mitchell, Donald W.	72
Mitchell, Mairin	72
Moerath, J. N.	72
Monosterev, N.	72
Montgery	72
Mordal, Jacques (C), see Auphan, Paul	5
Morison, Samuel Eliot	73
Moore, Don (C), see Fane, Francis D.	33
Muir, H. J.	73
La Neuville, C.	75
Newbolt, Henry	75
Newell, John S.	75
Newman, Al	75
Newman, James Ray	75
Nikolaev, Valerii Vasilevich	76
Nikolayev, B.	76
Nimitz, Chester W. (E), see Potter, E. B. (E & C)	83
Noalhat, H.	76
Nock, C. F.	76
Noel, Cdr. G. H. U	76
Nolan, E. H.	77
Normand, J. A.	77
Norris, Walter B. (C), see Krafft, Herman F.	54
Nutting, William Washburn	78
O'Hearn, Edward P.	78
Oliver, F. L	79
Oppenheim, Lassa F. L.	79
Ormsbey, Eugene	80
Orvin, Maxwell Clayton	80
Padgett, Harry E.	80
Palmer, Francis Ingraham	80
Palmer, Wayne F.	80

Name(s)	Page(s)
Pan-Se-Tcheng	80
Parsons, William Barclay	80
Patiens	80
Patterson, Andrew, Jr.	81
Paullin, Charles Oscar	81
Pavlovich, N. B.	81
Pawle, Gerald	81
"P.C." (NP)	81
Pelissier, Jean	81
Pendergast, Maurice (C), see Gibson, R. H.	39
Perlia, Sigmund Naumovich	81
Perry, Milton F.	81
Pesce, G. L.	81
Petrov, F. A. (C), see Samarov, A. A.	91
Pfankuchen, Llewellyn	81
Philipson, Coleman	81
Piehler, D. (C), see Hirshberg, N.	81
Piffera y Gallindo, Juan de la	81
Pinto, J. E. (C), see Lankford, B. W.	54
Piron, Cap. du Génie F. P. S.	82
Piterskii, Nikolai A. (E)	82
"P.L." (NP)	82
Pluddemann, Martin	82
Politovsky, E. S.	82
Polmar, Norman	82
Pope, Jennie Barnes (C), see Albion, Robert Greenhalgh	2
Porter, D. C.	82
Portlock, Ronald	83
Postan, M. M. (C), see Scott, J. D. or Hay, D.	83
Potter, E. B. (E & C), see Nimitz, C. W. (E)	83
Potter, Pitman B.	83
Powers, Robert D., Jr.	83
Pratt, Fletcher	83
Pratt, William V.	83
Puleston, W. C.	84
Pyke, G.	84
"Quarterdeck" (NP)	84
Rairden, Percy Wallace, Jr.	85
Ranson, M. A.	85
Reed, David	86
Reigart, J. Franklin	86
Reuter, Herbert C.	87

Name(s)	Page(s)
Reventlow, E. Graf	87
Reymond, P.	87
"R.G.E." (NP)	87
Richardson, J. B.	87
Riggs, Jerry	87
Riley, David R.	87
Robertson, W. W.	88
Robinson, Charles Napier (E), see Armstrong, Lt. G. E. (A)	88
Robinson, Reed A.	88
Robinson, S. S.	88
Rocholl, Erich	88
Roden, Ernest K.	88
Roebling Manufacturing	88
Rodgers, Hamp (C), see Gorman, Ron	88
Rodgers, Robert H.	88
Rogers, T. H. (C), see Chinn, W. K.	88
Rohan, Jack	88
Roland, Alex Frederick	88
Rollman, G.	88
von Romocki, S. J.	88
Roosevelt, Theodore	88
Roscoe, Theodore	89
Roskill, Stephen Wentworth	89
Rothbotham, W. B.	89
Rougerson, C.	89
Routledge, Robert	89
Rowe, O.	89
Rowland, Buford (C), see Boyd, William B.	89
Ruge, Friedrich O.	90
Rushmore, D. B.	90
Russell, William Howard (T), see Todleben (A)	105
Ryan, Cornelius	91
Ryan, L. S.	91
Saar, Charles W.	91
Samarov, A. A. (C), see Petrov, F. A.	91
Sanders, Harry	91
Sargent, Nathan	91
de Sarrepont, Major H. (Pseud. de Hennebert, Lt. Col. Eugene)	91
Saunders, M. G. (E)	91
Sauvaire-Jourdan, F.	91
Savage, Carlton	91
Savill, Merryn (T), see Galland, Adolf (A)	37
Scharf, J. Thomas	91
Scheer, Reinhard von	91, 92
Scheidnagel, D. Leopold	92

Name(s)	Page(s)
von Scheliha, Viktor Ernst Karl Rudolf	92
Schell, F. H.	92
Schofield, B. B.	92
Schoute-Vanneck, C. A. (C), see Gillett, J. K.	39
Schriebershofen, Max	92
Schull, Joseph	92
Schultz, J. H.	92
Schultz, M.	92
Schurman, D. M.	92
Schwarte, M.	92
Scott, James B. (E)	93
Scott, J. D. (C), see Postan, M. M. or Hay, D.	83
Scoville, Herbert, Jr.	93
Sears, James H.	93
Seawright, Murland W.	93
Semon, H. W.	94
Sendall, W. R.	94
Sessions, George Perry	94
Settle, Stuart Williston	94
Seuter, M. (A), see Bizot, M. (T)	10
Seves, Lt. de vais	94
Sewell, John Stephen	94
Shaw, Frances B.	94
Shearing, D.	94
Sheppard, William	95
Sheridan, M.	95
Sherwood, John	95
Shimanyak, B. A. (C), see Shulman, O. V.	95
Shulman, O. V. (C), see Shimanyak, B. A.	95
Sibley, N. W. (C), see Smith, F. E.	96
"Sic Fidem Teneo" (NP)	95
Sigsbee, Charles Dwight	95
Silas, Ellis	95
Simon, Leslie Earl	95
Sims, William Sowden	95
Siney, Marion C.	95
Skerrett, R. G.	96
Sleeman, Charles William	96
Smith, A. W.	96
Smith, C. J.	96
Smith, F. E. (C), see Sibley, N. W.	96
Smith, J. Bucknall	96
Smith, S. E.	96
Smythe, Augustine T.	96

Name(s)	Page(s)
Snow, Chester, R. (C), see Douglas, Harold G.	96
Sokol, Anthony Eugene	96
Sokol, Hans H.	96, 97
Soley, James Russell	97
Soulage, C. C.	97
Southall, Ivan	97
Sparks, Jared	97
Spindler, Arno	97, 98
Sprout, Harold and Margaret	98
Stadtlander, Gerd	98
Stafford, Edward P.	98
Stebbins, John	98
von Stengel, K.	98
Sterling, Y.	98
Stern, Philip von Doren	98
Steward, Harding	98
Stockton, Charles Herbert	98
Stokes, Anson Phelps	98
Stokes, Donald	98
Stone, Julius	98
Stotherd, Richard Hugh	99
"Strategicus" (NP)	99
Strauss, J.	99
Stroh, H.	99
Stroh, Henri (C), see Laubeuf, Alfred Maxime	54
Stryker, Lyal M.	99
Sturdee, Sir Fredrick C. D.	99
Suddath, Thomas H.	101
Sueter, M.	102
Sullivan, Barry W.	102
Sunderland, Archibald H.	102
Sundley, Sir Robert	102
"Taffrail" (NP), see Dorling, H. Taprell	28, 103
Tamura, Kyuzo	103
Taylor, Edmond	103
Taylor, William D., Captain, USN (Ret.)	104
Telberg, V. (E & P)	104
Thach, John W.	104
Thacher, James	104
Thomas, Hugh	104
Thomas, Lowell	104
Thomazi, Auguste	104
Thompson, R. W.	104
Thomson, David W.	104

Name(s)	Page(s)
Thorton, J. H., Jr. (C), see Karl, R. L.	52
Thursfield, James R. (C), see Clarke, Lt-Col Sir Georges S.	19
Thurston, Robert H.	104
Titherington, Richard H.	105
Todleben (A), see Russell, William Howard (T)	105
Tognelli, Vittorio (C), see Lupinacci, Pier Fillippo	58
Torry, John A. H.	108
Toudouze, G. G.	108
Touvet, G.	108
Townsend, A. O.	108
Trebesius, Ernst	108
Treuenfeld, R. von F.	108
Tryckare, Tre	108
Tucker, J. R. (C), see Gabbert, J. S. C. or Fothergill, J. W., Jr.	109
Tunstall, W. C. B. (C), see Keate, E. M.	52
Turner, Christopher R. (T), see Guichard, Louis (A)	42
Turner, John Frayn	109
Usborne, Cecil Vivian	110
Vance, Willis R.	110
Van der Veer, Norman R.	111
Van Duzer, E. M.	111
Van Nostrand, A. D.	111
Vergos, Lt de vaisseau	111
Veth, Kenneth L.	111
Vetkin, K. S.	111
Villers, G.	111
Virginskii, Viktor Semenovich	111
Vogel, Bertram	111
Vohtz, J. A.	111
Von der Porten, Edward P.	111
Vrangel, Ferdinand F.	112
"Vulcan" (NP)	112
Wade, H. A. L. H.	112
Wadlow, E. C.	112
Wages, C. J., Jr.	112
Wagner, Fredrick	112
Wainwright, Lawrence	112
Walke, Rear Admiral H.	112
Walker, J. Bernard	112
Walker, Sydney F.	112
Walsh, W. T.	113
Wandevelde, Capt Lt de Marine J. A.	113
Wang Shi Fu	113

Name(s)	Page(s)
Warneck, P. A.	113
Warner, Oscar C.	113
Warren, Harris Gaylord	113
Warshofsky, Fred	113
Washburn, S.	113
Washington, John	114
Webster, Charles (C), see Frankland, Noble	114
Welles, Gideon	114
Welling, William G.	114
Welliver, J. C.	114
Wells, L.	114
Westcott, Allan (C), see Alden, Carroll S.	2
Wester-Wemyss, R. E. W.	114
Wettern, Desmond	114
Wheddington, R. (C), see Crowther, James Gerald	23
Wheeler, J. B.	114
White, Fedoroff D.	114
White, Joe W.	114
White, W. L.	115
Whitehill, Walter M. (C), see King, Ernest J.	53
Whiteman, Marjorie M.	115
Wildrick, Meade	115
Williams, Frances L.	115
Wilmot, Chester (Reginald William Winchester Wilmot)	115
Wilson, George C.	115
Wilson, George Grafton (C)	115
Wilson, H. W.	116
Winn, Frank L.	116
Winterhalder, Theodore	116
Wisser, John P.	116
Wittmer, R.	116
Wolfram, E.	116
Wood, A. B.	116
Wood, Walter	116
Woodbury, David Oakes	116
Woodhouse, Henry	117
Woodward, David	117
Wulff, Olaf Richard (Vizeadmiral)	117
"X" (NP)	117
Yarham, E. R.	117
Yonge, Charles Duke	117
Young, Robert, Jr. (C), see Kirk, John	53
Zakharov, S. E.	117

REPORT DOCUMENTATION PAGE		READ INSTRUCTIONS BEFORE COMPLETING FORM
1. REPORT NUMBER NRC:MAC:2033	2. GOVT ACCESSION NO.	3. RECIPIENT'S CATALOG NUMBER
4. TITLE (and Subtitle) Historical Bibliography of Sea Mine Warfare		5. TYPE OF REPORT & PERIOD COVERED Bibliography
		6. PERFORMING ORG. REPORT NUMBER
7. AUTHOR(s) Andrew Patterson, Jr. Robert A. Winters		8. CONTRACT OR GRANT NUMBER(s) N00014-75-C-0258
9. PERFORMING ORGANIZATION NAME AND ADDRESS Mine Advisory Committee (now Naval Studies Board) National Research Council Washington, D.C. 20418		10. PROGRAM ELEMENT, PROJECT, TASK AREA & WORK UNIT NUMBERS
11. CONTROLLING OFFICE NAME AND ADDRESS Naval Studies Board National Research Council Washington, D.C. 20418		12. REPORT DATE January 1976
		13. NUMBER OF PAGES 139
14. MONITORING AGENCY NAME & ADDRESS(If different from Controlling Office)		15. SECURITY CLASS. (of this report) UNCLASSIFIED
		15a. DECLASSIFICATION/DOWNGRADING SCHEDULE

16. DISTRIBUTION STATEMENT (of this Report)

Publication has been authorized under terms of Contract N00014-75-C-0258 between the Office of Naval Research and the National Academy of Sciences. All inquiries concerning this report should be addressed to the Naval Studies Board, National Research Council, Washington, D.C. 20418. Copies of this report may also be obtained from the Defense Documentation Center.

17. DISTRIBUTION STATEMENT (of the abstract entered in Block 20, if different from Report)

18. SUPPLEMENTARY NOTES

19. KEY WORDS (Continue on reverse side if necessary and identify by block number)

Bibliography
Mines
Mine Countermeasures
Mine Warfare

20. ABSTRACT (Continue on reverse side if necessary and identify by block number)

REPORT DOCUMENTATION PAGE		READ INSTRUCTIONS BEFORE COMPLETING FORM
1. REPORT NUMBER NRC:MAC:2033	2. GOVT ACCESSION NO.	3. RECIPIENT'S CATALOG NUMBER
4. TITLE (and Subtitle) Historical Bibliography of Sea Mine Warfare		5. TYPE OF REPORT & PERIOD COVERED Bibliography
		6. PERFORMING ORG. REPORT NUMBER
7. AUTHOR(s) Andrew Patterson, Jr. Robert A. Winters		8. CONTRACT OR GRANT NUMBER(s) N00014-75-C-0258
9. PERFORMING ORGANIZATION NAME AND ADDRESS Mine Advisory Committee (now Naval Studies Board) National Research Council Washington, D.C. 20418		10. PROGRAM ELEMENT, PROJECT, TASK AREA & WORK UNIT NUMBERS
11. CONTROLLING OFFICE NAME AND ADDRESS Naval Studies Board National Research Council Washington, D.C. 20418		12. REPORT DATE January 1976
		13. NUMBER OF PAGES 139
14. MONITORING AGENCY NAME & ADDRESS(if different from Controlling Office)		15. SECURITY CLASS. (of this report) UNCLASSIFIED
		15a. DECLASSIFICATION/DOWNGRADING SCHEDULE

16. DISTRIBUTION STATEMENT (of this Report)

Publication has been authorized under terms of Contract N00014-75-C-0258 between the Office of Naval Research and the National Academy of Sciences. All inquiries concerning this report should be addressed to the Naval Studies Board, National Research Council, Washington, D.C. 20418. Copies of this report may also be obtained from the Defense Documentation Center.

17. DISTRIBUTION STATEMENT (of the abstract entered in Block 20, if different from Report)

18. SUPPLEMENTARY NOTES

19. KEY WORDS (Continue on reverse side if necessary and identify by block number)

Bibliography
Mines
Mine Countermeasures
Mine Warfare

20. ABSTRACT (Continue on reverse side if necessary and identify by block number)

DD FORM 1473 1 JAN 73 EDITION OF 1 NOV 65 IS OBSOLETE

www.ingramcontent.com/pod-product-compliance
Lightning Source LLC
Chambersburg PA
CBHW080557090426
42735CB00016B/3260